Science and Technology Concepts–Secondary™

Exploring
the Nature
of Light

Student Guide

National Science Resources Center

The National Science Resources Center (NSRC) is operated by the Smithsonian Institution to improve the teaching of science in the nation's schools. The NSRC disseminates information about exemplary teaching resources, develops curriculum materials, and conducts outreach programs of leadership development and technical assistance to help school districts implement inquiry-centered science programs.

Smithsonian Institution

The Smithsonian Institution was created by an act of Congress in 1846 "for the increase and diffusion of knowledge..." This independent federal establishment is the world's largest museum complex and is responsible for public and scholarly activities, exhibitions, and research projects nationwide and overseas. Among the objectives of the Smithsonian is the application of its unique resources to enhance elementary and secondary education.

STC Program™ Project Sponsors

National Science Foundation

Bristol-Meyers Squibb Foundation

Dow Chemical Company

DuPont Company

Hewlett-Packard Company

The Robert Wood Johnson Foundation

Carolina Biological Supply Company

Science and Technology Concepts–Secondary™

Exploring the Nature of Light

Student Guide

The STC Program™

Smithsonian Institution

National Science Resources Center

www.carolinacurriculum.com

Published by Carolina Biological Supply Company
Burlington, North Carolina

NOTICE This material is based upon work supported by the National Science Foundation under Grant No. ESI-9618091. Any opinions, findings, and conclusions or recommendations expressed in this material are those of the authors and do not necessarily reflect views of the National Science Foundation or the Smithsonian Institution.

This project was supported, in part, by the **National Science Foundation**.
Opinions expressed are those of the authors and not necessarily those of the foundation.

ISBN 978-1-4350-0718-5

Published by Carolina Biological Supply Company, 2700 York Road, Burlington, NC 27215. Call toll free 1-800-334-5551.

Science and Technology Concepts—Secondary™
Exploring the Nature of Light

The following revision was based on the STC/MS™ module *Light*.

Developer
Dane J. Toler

Scientific Reviewer
Ramon E. Lopez
Professor of Physics
University of Texas at Arlington

Illustrator
John Norton

Editor
Linda Harteker

Photo Research
Jane Martin
Devin Reese

National Science Resources Center Staff

Executive Director
Sally Goetz Shuler

Program Specialist/Revision Manager
Elizabeth Klemick

Contractor, Curriculum Research and Development
Devin Reese

Publications Graphics Specialist
Heidi M. Kupke

Carolina Biological Supply Company Staff

Director of Product and Development
Cindy Morgan

Marketing Manager, STC–Secondary™
Jeff Frates

Curriculum Editors
Lauren Goldsmith
Gary Metheny

Managing Editor, Curriculum Materials
Cindy Vines Bright

Publications Designers
Trey Foster
Charles Thacker
Greg Willette

Science and Technology Concepts for Middle Schools™
Light
Original Publication

Module Development Staff

Developer/Writer
David Marsland

Science Advisor
Alan Migdall

Contributing Writer
Lynda DeWitt

Illustrator
John Norton

STC/MS™ Project Staff

Principal Investigator
Sally Goetz Schuler, Executive Director, NSRC

Project Director
Kitty Lou Smith

Curriculum Developers
David Marsland
Henry Milne
Carol O'Donnell
Dane J. Toler

Illustration Coordinator
Max-Karl Winkler

Photo Editor
Christine Hauser

Graphic Designer
Heidi M. Kupke

STC/MS™ Project Advisors

Dr. David Branning, Department of Physics, University of Illinois at Urbana—Champaign

James Bickel, Teacher and Instructional Services, Minneapolis Public Schools

Caren Falascino, Fort Couch Middle School, Upper St. Clair

Dr. Jay Grinstead, National Institute of Standards and Technology, Optical Technology Division

Mr. Leonard Hanssen, National Institute of Standards and Technology, Optical Technology Division

Dr. Jack Hehn, Manager, Division of Education, American Institute of Physics

Mr. Paul Klein, Naval Research Laboratory

Mr. Robert Latham, Thomas Jefferson Secondary School for Science and Technology, Fairfax County, Virginia

Dr. John Layman, Professor Emeritus of Education and Physics, University of Maryland

Ms. Yvonne Mah, Montgomery County Public Schools, Maryland

Dr. Alan Migdall

Mr. Mike Isley, Carolina Biological Supply Company

Dr. Howard Yoon, Project Leader, Spectroradiometry, National Institute of Standards and Technology, Optical Technology Division

Acknowledgments

The National Science Resources Center gratefully acknowledges the following individuals and school systems for their assistance with the national field-testing of *Light:*

California

Site Coordinator
Leona Lallier
El Centro Elementary School District

El Centro Elementary School District
El Centro
John Lazarcik, Teacher
Kennedy Middle School

Calexico Unified School District
Calexico
Yolanda Guerrero, Teacher
William Moreno Junior High School

Holtville Unified School District
Holtville
Richard Sanchez, Teacher
Holtville Middle School

Maryland

Montgomery County Public Schools
Yvonne Mah, Teacher
Shady Grove Middle School, Gaithersburg

Minnesota

Site Coordinator
James Bickel, Teacher and Instructional Services
Minneapolis Public Schools

Minneapolis Public Schools
Minneapolis
John Roper-Batker, Teacher
Seward Montessori School

Tracey Schultz, Teacher
Franklin Middle School

Diane Weiher, Teacher
Franklin Middle School

North Carolina

Alamance-Burlington School System
Burlington
Elizabeth Thornburg, Teacher
Woodlawn Middle School, Mebane

Pennsylvania

Site Coordinator
Jim Smoyer
Allegheny Schools Science Education and
Technology (ASSET), Pittsburgh

Upper Saint Clair School District
Upper Saint Clair
Nelson Early, Teacher
Fort Couch Middle School

Caren Falascino, Teacher
Fort Couch Middle School

Northgate School District
Pittsburgh
Frank Nesko, Teacher
Northgate Junior and Senior High School

Tennessee

Site Coordinator
Jimmie Lou Lee, Center for Excellence for
Research and Policy on Basic Skills
Tennessee State University, Nashville

Williamson County Schools
Barbara Duke, Teacher
Grassland Middle School, Franklin

Metro Nashville Public Schools
Nashville
Kathy Lee, Teacher
Martin Luther King Magnet School

Sumner County Schools
Shelli White, Teacher
Rucker Stewart Middle School, Gallatin

The NSRC appreciates the contribution of its
STC/MS project evaluation consultants—

Program Evaluation Research Group (PERG), Lesley College

Sabra Lee
Researcher, PERG

Center for the Study of Testing, Evaluation,
and Education Policy (CSTEEP), Boston College

Joseph Pedulla
Director, CSTEEP

Preface

Community leaders and state and local school officials across the country are recognizing the need to implement science education programs consistent with the National Science Education Standards to attain the important national goal of scientific literacy for all students in the 21st century. The Standards present a bold vision of science education. They identify what students at various levels should know and be able to do. They also emphasize the importance of transforming the science curriculum to enable students to engage actively in scientific inquiry as a way to develop conceptual understanding as well as problem-solving skills.

The development of effective standards-based, inquiry-centered curriculum materials is a key step in achieving scientific literacy. The National Science Resources Center (NSRC) has responded to this challenge through Science and Technology Concepts–Secondary™. Prior to the development of these materials, there were very few science curriculum resources for secondary students that embodied scientific inquiry and hands-on learning. With the publication of STC–Secondary™, schools will have a rich set of curriculum resources to fill this need.

Since its founding in 1985, the NSRC has made many significant contributions to the goal of achieving scientific literacy for all students. In addition to developing Science and Technology Concepts–Elementary™—an inquiry-centered science curriculum for grades K through 6—the NSRC has been active in disseminating information on science teaching resources, preparing school district leaders to spearhead science education reform, and providing technical assistance to school districts. These programs have had a significant impact on science education throughout the country. The transformation of science education is a challenging task that will continue to require the kind of strategic thinking and insistence on excellence that the NSRC has demonstrated in all of its curriculum development and outreach programs. The Smithsonian Institution, our sponsoring organization, takes great pride in the publication of this exciting new science program for secondary students.

Letter to the Students

Smithsonian Institution
National Science Resources Center

Dear Student,

The National Science Resources Center's (NSRC) mission is to improve the learning and teaching of science for K-12 students. As an organization of the Smithsonian Institution, the NSRC is dedicated to the establishment of effective science programs for all students. To contribute to that goal, the NSRC has developed and published two comprehensive, research-based science curriculum programs: Science and Technology Concepts-Elementary™ and Science and Technology Concepts-Secondary™.

By using the STC-Secondary™ curriculum materials, we know that you will build an understanding of important concepts in life, earth, and physical sciences; learn critical-thinking skills; and develop positive attitudes toward science and technology. The National Science Education Standards state that all secondary students "...should be provided opportunities to engage in full and partial inquiries.... With an appropriate curriculum and adequate instruction, ... students can develop the skills of investigation and the understanding that scientific inquiry is guided by knowledge, observations, ideas, and questions."

STC-Secondary also addresses the national technology standards published by the International Technology Education Association. Informed by research and guided by standards, the design of the STC-Secondary units addresses four critical goals:

- Use of effective student and teacher assessment strategies to improve learning and teaching

- Integration of literacy into the learning of science by giving students the lens of language to focus and clarify their thinking and activities

- Enhanced learning using new technologies to help students visualize processes and relationships that are normally invisible or difficult to understand

- Incorporation of strategies to actively engage parents to support the learning process

We hope that by using the STC-Secondary curriculum you will expand your interest, curiosity, and understanding about the world around you. We welcome comments from students and teachers about their experiences with the STC-Secondary program materials.

Sally Goetz Shuler
Executive Director
National Science Resources Center

Navigating an STC–Secondary™ Student Guide

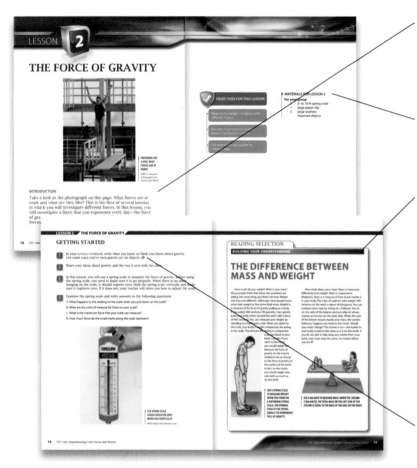

INTRODUCTION
This short paragraph helps get you interested about the upcoming inquiries.

MATERIALS
This helps you get organized and prepare for your inquiries.

READING SELECTION:
BUILDING YOUR UNDERSTANDING
These reading selections are part of the lesson, and give you information about the topic or concept you are exploring.

NOTEBOOK ICON
During the course of an inquiry, you'll record data in different ways. This icon lets you know to record in your science notebook. Student sheets are called out when you're to write there. You may go back and forth between your notebook and a student sheet. Watch carefully for the icon throughout the procedure.

SAFETY TIPS
Safety in the science classroom is very important. Tips throughout the student guide will help you to practice safe techniques while conducting investigations. It is very important to read and follow all safety tips.

PROCEDURE

This tells you what to do. Sometimes the steps are very specific, and sometimes they guide you to come up with your own investigation and ways to record data.

REFLECTING ON WHAT YOU'VE DONE

These questions help you think about what you've learned during the lesson's inquiries, apply them to different situations, and generate new questions. Often you'll discuss your ideas with the class.

READING SELECTION: EXTENDING YOUR KNOWLEDGE

These reading selections come after the lesson, and show new ways that the topic or concept you learned about during the lesson can be applied, often in real-world situations.

GLOSSARY

Here you can find scientific terms defined.

INDEX

Locate specific information within the student guide using the index.

Contents

CONTENTS

THINKING ABOUT LIGHT

▶ **THESE STUDENTS ARE INVESTIGATING LIGHT AT THEIR LOCAL SCIENCE CENTER.**

PHOTO: Amy Snyder, © Exploratorium, www.exploratorium.edu

INTRODUCTION

How much do you know about light? When do we encounter light? How do we use it and how does light behave? This first lesson is designed to get you thinking about these questions. Working collaboratively, you will conduct a series of short inquiries. You will make and record observations and discuss what you think is happening in each inquiry. If you have questions about what is happening, you will write them down. Later in the lesson, you will share your observations, ideas, and questions with your classmates. You will revisit your ideas and questions about light as you proceed through the unit.

OBJECTIVES FOR THIS LESSON

Conduct a series of short inquiries about light.

Make and record observations.

Discuss your observations and ideas about what is happening.

Identify and share questions you might have about light.

MATERIALS FOR LESSON 1

For you

1　copy of Student Sheet 1: Thinking About Light

For your group

1　transparent cup (three-fourths filled with water)

1　aluminum nail

2　index cards

1　marker

GETTING STARTED

1 Read the introduction and listen as your teacher introduces the unit.

2 Look at the nail and then stand it in the cup of water. Look at the nail in the water from a variety of different angles. What do you observe?

3 Discuss what you have observed with your group. You will be asked to share your observations with the class.

4 Record the class's observations (write or draw them) in the second column of Table 1 on Student Sheet 1: Thinking About Light.

5 Can you explain what is happening? Share your ideas with the class. Record some of the class's ideas in the third column of Table 1.

6 Think of some questions that arise from your observations. Share these questions with the class.

7 Write any questions the class has about this exercise in the fourth column of Table 1. Clean up your cup of water and return your nail to your teacher.

PROCEDURE FOR THE CIRCUIT OF INQUIRIES

You are now ready to conduct the inquiries in this lesson. Just as in "Getting Started," you will discuss and record your observations, your ideas about what is happening, and your questions in Table 1 on Student Sheet 1. Here are a few general instructions:

1 You will conduct inquiries with your group at four different stations. The inquiries are numbered from 1.1-1.4 (or 1.1A-1.4A). Each inquiry has instructions you need to follow and questions designed to help you make observations and think about what is happening.

2 Your teacher will tell your group at which station to begin.

3 You will have 4-5 minutes to perform the inquiry and make observations. Record your observations, ideas, and questions for each inquiry in the correct row of Table 1.

4 When your teacher calls time, make sure you leave the station as you found it before moving on to the next inquiry in the sequence.

5 When you have completed all four inquiries, return to your desk.

6 Record any additional questions you have about what you observed.

LAMP LIGHT

PROCEDURE

1 Slowly slide the dimmer switch up and down (see Figure 1.1). Record your observations.

2 What do you think is happening? Discuss your ideas with your group before you write them down.

3 Record any questions you have about what you observed.

4 Switch off the lamp.

SAFETY TIP

Do not touch the lightbulb. It gets hot and may burn your fingers!

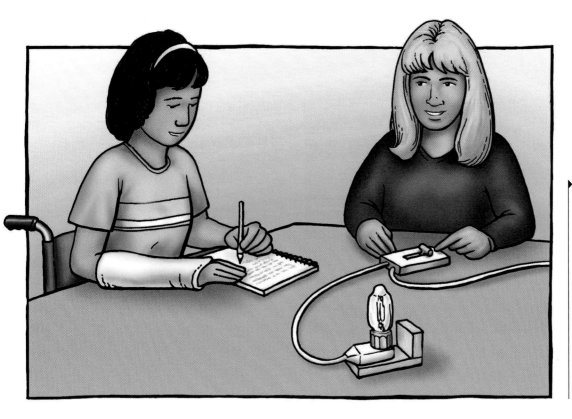

▶ CAN YOU DESCRIBE WHAT HAPPENS INSIDE THE LIGHTBULB WHEN YOU SLIDE THE DIMMER SWITCH UP AND DOWN? (NOTE: THE DIMMER SWITCH YOU ARE USING MAY DIFFER SLIGHTLY FROM THE ONE SHOWN.)
FIGURE **1.1**

INQUIRY **1.2**

THE RADIOMETER

PROCEDURE

1 Switch on the flashlight. Point the beam at the radiometer. Record what you observe.

2 What effect does moving the flashlight nearer to and farther from the radiometer have on the radiometer? (See Figure 1.2.) Record what you think is happening.

3 Record any questions you have about what you observed.

WHAT EFFECT DOES MOVING THE FLASHLIGHT NEARER TO AND FARTHER FROM THE RADIOMETER HAVE ON THE RADIOMETER?
FIGURE **1.2**

COLORED LIGHTBULBS

PROCEDURE

SAFETY TIP

The lightbulbs are very hot. Do not touch the lightbulbs with your hand or the paper.

1 Switch on the power strip with the three colored lightbulbs.

2 Hold the piece of paper above the lightbulbs and look at the paper (see Figure 1.3).

3 Being careful not to touch the lightbulbs, place one hand between the paper and the lightbulbs.

4 Describe what you see on the paper.

5 Try to explain to your group what you observe. Record your ideas, explanations, and any questions you have about this inquiry.

6 Switch off the strip of lightbulbs.

▶ **HOLD THE PAPER ABOVE THE COLORED LIGHTBULBS.**
FIGURE **1.3**

INQUIRY 1.4

LOOKING BEHIND

PROCEDURE

1 Hold the mirror in front of you while a group member stands behind you. Move the mirror so that you can see the person behind you. Record your observations.

2 Try drawing a sketch in the third column of Table 1 that explains how you can see someone behind you.

3 Record any questions you have about what you observed.

REFLECTING
ON WHAT
YOU'VE DONE

1 Share your observations, ideas, and explanations with your grou[p]. Compare the questions you have generated.

2 From these questions, select the two questions that your group members agree they would mos[t] like to be able to answer. Work with your group to improve the wording of these two questions.

A. Write the questions on your student sheet. Be sure to indicate which inquiry each question came from.

3 Your teacher will give your grou[p] two index cards and a marker. Us[e] the marker to write one questio[n] on each card. Write clearly and include the number of the inqui[ry] your question relates to at the top of the card (see Figure 1.4).

4 Your teacher will lead a review o[f] the class's observations, ideas, a[nd] questions. Here are some things you might like to think about during this review:

A. Have some groups asked simil[ar] questions? Are some versions of these questions better than others?

B. Can the questions be organize[d] into groups or by topic? Can you suggest names for these groups or topics?

▶ DISCUSS YOUR QUESTIONS ABOUT THE INQUIRIES WITH YOUR GROUP. IDENTIFY YOUR GROUP'S TWO QUESTIONS AND WRITE ONE QUESTION ON EACH OF THE INDEX CARDS AS SHOWN.
FIGURE **1.4**

Inquiry #1.1

How does the dimmer switch work?

Using and Studying Light

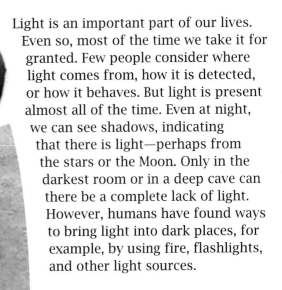

Light is an important part of our lives. Even so, most of the time we take it for granted. Few people consider where light comes from, how it is detected, or how it behaves. But light is present almost all of the time. Even at night, we can see shadows, indicating that there is light—perhaps from the stars or the Moon. Only in the darkest room or in a deep cave can there be a complete lack of light. However, humans have found ways to bring light into dark places, for example, by using fire, flashlights, and other light sources.

▶ HUMANS USED FIRE FOR THOUSANDS OF YEARS TO PROVIDE LIGHT AND HEAT. THESE MASAI TRIBESMAN ARE MAKING A FIRE USING STICKS, DRIED WOOD, AND ELEPHANT DUNG. HOW DOES A FIRE LIKE THE ONE THEY ARE MAKING PRODUCE LIGHT?

PHOTO: dougwoods/creativecommons.org

READING SELECTION
EXTENDING YOUR KNOWLEDGE

▶ THESE FLOWERS OPEN DURING THE DAY AND CLOSE AT NIGHT. TO DO THIS, THE PLANT MUST HAVE SOME WAY TO DETECT LIGHT. HOW DOES THE PLANT DETECT LIGHT? HOW DO WE DETECT LIGHT?

PHOTO: Tony Hisgett/ creativecommons.org

SENSING LIGHT

Light is so important to living things that almost all organisms can detect light in some way. For instance, plants grow in such a way as to bend toward light. Microbes sometimes move either toward or away from light.

Most animals have special light detectors. Humans have two light detectors: their eyes. Without light we couldn't use one of our five senses—sight. We would have no way to see the world around us. No one could enjoy the work of great painters who express themselves using shape and color.

▶ LIGHT CAN BE USED TO ENTERTAIN. AT A CONCERT, LOUD MUSIC AND FLASHING LIGHTS CREATE AN ATMOSPHERE OF FUN AND ENTERTAINMENT.

PHOTO: mattgarber/creativecommons.org

► WITHOUT A LIGHT SOURCE, THESE CAVERS WOULD SEE NOTHING OF THIS IMMENSE CAVERN.

PHOTO: Courtesy of Gary Berdeaux

READING SELECTION
EXTENDING YOUR KNOWLEDGE

LIGHT PIONEERS

Scientists and artists have studied light for thousands of years.

▶ LEONARDO DA VINCI (1452–1519) HAD ONE OF THE MOST CURIOUS MINDS IN HISTORY. THIS ITALIAN ARTIST, SCIENTIST, AND ENGINEER OBSERVED LIGHT AND ASKED HIMSELF MANY QUESTIONS ABOUT IT. HE THEN APPLIED THE RESULTS OF HIS INQUIRIES INTO THE NATURE OF LIGHT, REFLECTIONS, AND SHADOWS TO HIS GREAT WORKS OF ART.

PHOTO: Library of Congress, Prints & Photographs Division, LC-USZ62-111797

▶ THE SCIENTIST ALBERT EINSTEIN (1879–1955) IS PERHAPS THE MOST FAMOUS OF THE MANY SCIENTISTS WHO HAVE CONTRIBUTED TO A BETTER UNDERSTANDING OF THE NATURE OF LIGHT OVER THE PAST HUNDRED YEARS.

PHOTO: Library of Congress, Prints & Photographs Division, LC-USZ62-60242

Without light we couldn't use some types of technology to learn, entertain, and communicate. Photographers used light to produce the pictures you see in this book. Movies, television shows, and computer games use light to entertain us. People even listen to music by using laser light in CD players. Make a phone call or send an e-mail or instant message to your friends, and chances are you are communicating through tiny glass fiber-optic cables that carry your message as pulses of light.

LIGHT AND SCIENCE

As scientists work to understand the natural world, they use light to make observations. They may use special instruments that use light to assist them in this work. For example, astronomers use telescopes to observe the night sky. They use other instruments to analyze light from distant stars and galaxies to determine the substances that compose them. Biologists use microscopes to uncover the hidden workings of cells. But before scientists could construct these

▶ OUR KNOWLEDGE OF OPTICS HAS ENABLED US TO BUILD GIANT TELESCOPES THAT HELP US SEE DEEP INTO THE UNIVERSE. THIS TELESCOPE IS AT GRIFFITH OBSERVATORY IN CALIFORNIA. THE SAME KNOWLEDGE ALLOWS US TO LOOK AT THE WORLD OF THE VERY SMALL, THE MICROSCOPIC.

PHOTO (right): DoD photo by Petty Officer 2nd Class Ryan C. McGinley, U.S. Navy
PHOTO (below): Clinton Steeds/creativecommons.org

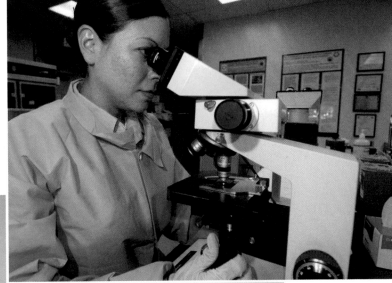

modern scientific tools, they had to learn about the nature and behavior of light. This study of light is called optics.

Interest in optics can be traced back thousands of years—even before recorded history. The first book about optics was written more than two thousand years ago. Early questions about light may have included the following: Why can reflections be seen on the surface of water? Why does sunlight produce fire when it passes through specially shaped polished crystals? Perhaps the most difficult question may have been, What is light? This question continues to preoccupy even modern scientists. ■

DISCUSSION QUESTIONS

1. Think of 10 ways in which you depend on light in your everyday life.

2. Why is the study of optics important?

The Sun: A Source of Light, Myth, and Tradition

Each dawn, not far from the shores of the Bay of Bengal, India, the first rays of the rising Sun strike the 13th-century sun temple of Konark. The temple was built in the form of a giant 24-wheeled chariot drawn by seven horses to honor the sun god Surya. According to legend, only such a magnificent chariot was fit to carry Surya across the sky in his dawn-to-dusk journey. Six hours later, the Sun's early morning rays touch the stones of Stonehenge in England. Throughout history, people from all over the world have worshipped the Sun, our closest star. Ancient Britons built this giant astronomical calendar and center for sun worship four thousand years ago. They recognized it as the source of the light and heat so essential for life. No wonder many cultures gave the Sun the status of a god.

The early Egyptians built elaborate temples to their sun gods, Re and Aton. The ancient Greeks also had sun gods—first Helios and then Apollo. The Greeks also explained the Sun's apparent movement across the sky in terms of a golden chariot, light for both humans and gods. In addition to worshipping sun gods, some peoples—like the Japanese—believed that their rulers were descended from sun gods.

▶ THE KONARK TEMPLE IN INDIA WAS BUILT IN THE FORM OF A CHARIOT. IN LEGEND, THIS HORSE-DRAWN CHARIOT CARRIES THE SUN GOD SURYA ACROSS THE SKY.

PHOTO: Nomad Tales/ creativecommons.org

▶ ANCIENT BRITONS BUILT STONEHENGE AS AN ASTRONOMICAL CALENDAR AND A PLACE OF SUN WORSHIP. SOME OF ITS GIANT STONES WERE DRAGGED OVER 160 KILOMETERS (100 MILES) TO THE SITE. THE STONES ARE ARRANGED SO THAT ON A MID-SUMMER'S DAY THE RAYS OF THE RISING SUN SHINE INTO THE CENTER OF THE MONUMENT.

PHOTO: javajones/creativecommons.org

▶ A WHEEL FROM A TEMPLE CHARIOT FIT FOR A SUN GOD.

PHOTO: Nomad Tales/creativecommons.org

CELEBRATING THE SUN

Have you ever thought about where the word "Sunday" came from? In A.D. 313, the Roman emperor Constantine became a Christian. He then changed the day for worshipping the Roman sun god, Sol Victis, into a day for worship for Christians—Sunday. Even celebrations on December 25 began as a day of sun worship. Originally celebrated as the Feast of Sun in India, this celebration became Christmas in the West.

▶ UNTIL 1945, THE JAPANESE ROYAL FAMILY TRACED ITS DESCENT FROM THE SUN GODDESS AMATERASU OMIKAMI, A GODDESS OF ONE OF JAPAN'S OLDEST RELIGIONS, SHINTO. HERE A WOMAN PERFORMS A DANCE TO DRAW AMATERASU OUT OF HER CAVE.

PHOTO: Library of Congress, Prints & Photographs Division, LC-DIG-jpd-01968

Celebration of the Sun also was common among the early inhabitants of the Americas. The Aztecs, Incas, and Maya recognized the Sun as a deity. They staged elaborate rituals and sacrifices in temples created for these gods.

Some of the native peoples of the North American plains still hold a renewal ceremony, called the Sun Dance, in spring or early summer. This four-day event of rituals and dances celebrates the Sun and the forces of nature.

The Sun has long been important to the culture and traditions of the world's peoples. Modern scientists explain the Sun not as a god, but as a giant ball of gas—about 1,400,000 kilometers (about 870,000 miles) in diameter—that releases light and heat from the nuclear reactions that occur inside it. ■

JUST AS THE STATUE OF LIBERTY STANDS AT THE GATEWAY TO NEW YORK, SO A STATUE OF THE SUN GOD HELIOS GUARDED THE ANCIENT HARBOR OF RHODES, A GREEK ISLAND. THE STATUE, KNOWN AS THE COLOSSUS OF RHODES, IS CONSIDERED ONE OF THE SEVEN WONDERS OF THE ANCIENT WORLD. IT STOOD MORE THAN 30 METERS (98.4 FEET) TALL AND WAS MADE FROM BRONZE MELTED DOWN FROM THE WEAPONS OF A DEFEATED ENEMY.

THE SURFACE OF THE SUN IS VERY HOT AND RELEASES A LOT OF LIGHT. ARE LIGHT AND HEAT RELATED?

PHOTO: NOAA

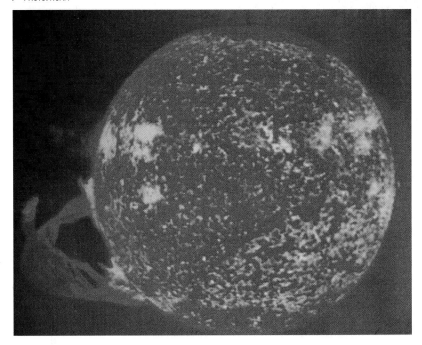

 DISCUSSION QUESTIONS

1. Why do you think that various cultures have worshipped the Sun as a god?

2. Research several examples of sun worship other than those mentioned in the reading selection.

WHERE DOES LIGHT COME FROM?

INTRODUCTION

Everyone knows that light exists. But what is light? Where does light come from, and how is it made? From your own observations, you know that some objects make light. The most important of these is the Sun. Do you know how the Sun and other objects make light? In this lesson, you will try to answer some of these questions. You will identify different sources of light and then conduct an inquiry to examine two of these sources—a flashlight and a lit candle—in more detail. You will discuss the nature of light and some processes by which it is made.

▶ WE USE ARTIFICIAL LIGHT SOURCES TO HELP US TO SEE. HOW DO LIGHT SOURCES LIKE FLASHLIGHTS PRODUCE LIGHT? WHERE DOES LIGHT COME FROM?

PHOTO: DoD photo by A1C Isaac G.L. Freeman, USAF

OBJECTIVES FOR THIS LESSON

Share ideas on the nature of light.

Identify different sources of light.

Investigate some sources of light.

Discuss light as a form of energy.

Discuss a range of energy transformations.

Investigate and discuss the transformation of light into other forms of energy.

▶ MATERIALS FOR LESSON 2

For you

1   copy of Student Sheet 2.2: How Is Light Produced?

1   pair of safety goggles

For your group

1   flashlight

2   D-cell batteries

1   petri dish (lid or base)

1   tea candle

GETTING STARTED

INQUIRY 2.1

1 In your science notebook, record your own ideas about the following questions: ✏

   A. What do you think light is?

   B. What do you think light is not?

   C. Where does the light in the classroom come from?

2 Be prepared to share your ideas with the class.

IDENTIFYING SOURCES OF LIGHT

PROCEDURE

1 You have identified a source, or some sources, of light within your classroom. Working with your group, think of other sources of light. List them in your science notebook. ✏

2 Your teacher will ask you to contribute to a class brainstorm on different light sources. Your teacher will record the class's ideas on a concept map.

3 As the concept map develops, record it in your science notebook.

▶ **IN THIS LESSON, YOU WILL BE THINKING ABOUT SOURCES OF LIGHT.**

PHOTO: jennifer chong/creativecommons.org

HOW IS LIGHT PRODUCED?

PROCEDURE

1 One member of your group should collect the materials as directed by your teacher. Record your responses on Student Sheet 2.2: How Is Light Produced?

2 Working with your group, examine the flashlight (see Figure 2.1). You may take it apart if you wish, but be careful not to break it. Do not dismantle the switch. As you examine the flashlight, answer the following questions. Use words and/or diagrams to record your results and ideas under Step 1 on Student Sheet 2.2.

A. Exactly where does the flashlight release its light?

B. Where does this light come from?

C. Where does the light go?

D. Is anything produced in addition to light?

E. Try to explain how the flashlight makes light.

SAFETY TIPS

Put on your safety goggles before proceeding with the inquiry.

Your teacher may give you a book of matches to use to light the candle. Use caution when lighting the match.

Always strike a match away from you and other students, and be very careful not to burn your fingers.

Do not play with the matches.

Avoid breathing in the fumes from the match.

▶ TRY TO FIND OUT HOW A FLASHLIGHT PRODUCES LIGHT.
FIGURE **2.1**

Inquiry 2.2 continued

3 Reassemble the flashlight. Check to see that it works.

4 Place the candle on the petri dish. Use a match (or ask your teacher) to ignite the candle. Observe what happens as the match is struck and ignites the candle.

5 Think about the following questions, and then record your answers under Step 2 on Student Sheet 2.2:

A. What was produced when you struck the match?

B. Where did it come from?

C. How does the candle make light?

D. Is anything else produced in addition to light?

E. What happens to the candle as it makes light?

6 Extinguish your candle. Return all the materials as directed by your teacher.

7 Participate in a class discussion about how the candle produces light.

TRANSFORMING ENERGY

Different light sources make light in different ways. But all light sources have something in common. They produce light as a result of energy transformations. Energy can take many forms. (Light is just one form of energy. Heat energy, electrical energy, movement energy, and chemical energy—stored in food or fuel—are also forms of energy. All forms of energy can do work—they can make things move.)

Energy can also be transformed from one form of energy into another form. For example, switching on the lights in your classroom allows electrical energy to enter the fluorescent tubes, where some of this electrical energy is transformed into light.

When an energy transformation takes place, usually more than one form of energy is produced. Switching on the classroom lights releases light energy. But, if you touch a lit fluorescent tube, you will also notice it is slightly warm. Heat is almost always one of the forms of energy released during an energy transformation. The energy transformation occurring inside the tube could be expressed as follows:

ELECTRICAL ENERGY → LIGHT and HEAT

Was there any evidence that heat was released during the energy transformations you observed in the flashlight, match, or candle? ■

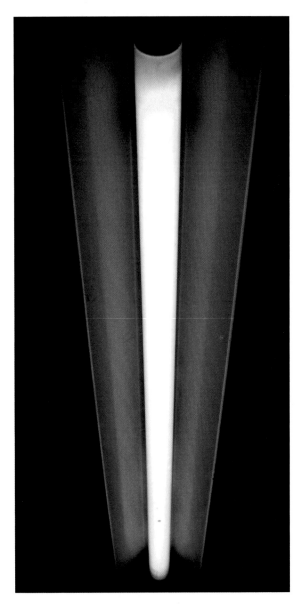

IN A FLUORESCENT TUBE, ELECTRICAL ENERGY PASSES INTO MERCURY VAPOR INSIDE THE TUBE AND EXCITES THE MERCURY VAPOR. THE EXCITED VAPOR MAKES THE WHITE COATING ON THE INSIDE OF THE TUBE GLOW.

PHOTO: Josep Mª Rosell/creativecommons.org

WHERE'S THE ENERGY?

▶ IDENTIFY THE FORMS OF ENERGY YOU SEE IN THIS SCENE. MARK
AND WRITE THE NAMES OF THESE FORMS OF ENERGY ON OR
AROUND THE EDGE OF THE PICTURE ON STUDENT SHEET 2.2.
FIGURE **2.2**

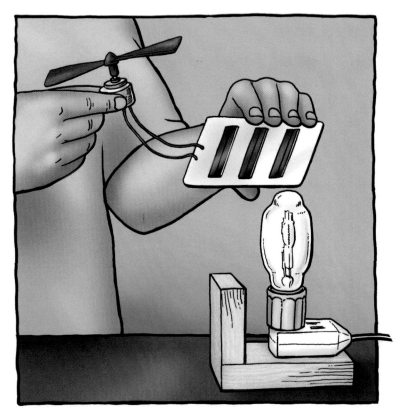

▶ WHAT HAPPENS TO THE MOTOR WHEN LIGHT STRIKES THE SOLAR PANEL? WHAT ENERGY TRANSFORMATIONS ARE INVOLVED?
FIGURE **2.3**

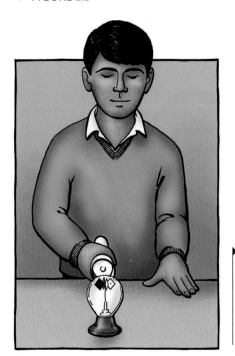

▶ WHAT HAPPENS WHEN A FLASHLIGHT SHINES ON THE RADIOMETER? WHAT ENERGY TRANSFORMATIONS ARE INVOLVED?
FIGURE **2.4**

REFLECTING
ON WHAT
YOU'VE DONE

1. Read "Transforming Energy" on page 23.

2. Discuss the following questions with your group. Record your responses in your science notebook.

   A. Was light stored in any of the items you examined?

   B. What had to happen for light to be produced?

   C. What energy transformations took place in the flashlight and the candle?

3. Look at the scene shown in Figure 2.2. (This is the same scene as the "Where's the Energy?" picture on Student Sheet 2.2.) Answer the questions under Step 3 on your Student Sheet.

4. As a form of energy, it should be possible to transform light energy into other forms. Watch as your teacher shows you two examples of such a transformation (see Figures 2.3 and 2.4).

   A. What energy transformations are taking place with the solar panel system? With the radiometer? Record your ideas in your science notebook.

   B. Write a short paragraph outlining the evidence that light is a form of energy.

5. Look at the question bank generated in Lesson 1. Can you answer any of the questions using what you discovered during this lesson?

READING SELECTION

EXTENDING YOUR KNOWLEDGE

SOURCES
of Light

Light sources—objects that make their own light, such as lightbulbs, the Sun, or candles—are said to be luminous. Some examples of luminous objects are shown here. Can you think of others? ■

▶ IN A TRADITIONAL LIGHTBULB, THE FILAMENT—USUALLY A PIECE OF THIN TUNGSTEN—GLOWS WHEN ELECTRICAL ENERGY PASSES THROUGH IT. THE FILAMENT DOES NOT BURN UP BECAUSE IT IS ENCLOSED IN A GLASS BULB CONTAINING A NON-REACTIVE GAS, SUCH AS ARGON. IN ADDITION TO LIGHT, THESE LIGHTBULBS PRODUCE QUITE A LOT OF WASTE HEAT—HEAT THAT IS NOT USED. THIS WASTED HEAT MEANS THESE LIGHTBULBS ARE NOT ENERGY EFFICIENT. FLUORESCENT LIGHTBULBS, WHICH PRODUCE FAR LESS WASTE HEAT, ARE MORE ENERGY EFFICIENT.

PHOTO: Clark Gregor/*creativecommons.org*

▶ LIGHT-EMITTING DIODES (LEDS)—LIKE THE ONES THAT MAKE UP THE NUMBERS IN THIS CLOCK—PRODUCE LIGHT FROM ELECTRICAL ENERGY. BECAUSE THEY PRODUCE ALMOST NO HEAT, THEY ARE MUCH MORE EFFICIENT AT PRODUCING LIGHT THAN TRADITIONAL LIGHTBULBS.

PHOTO: Jeff McAdams, Photographer, Courtesy of Carolina Biological Supply Company

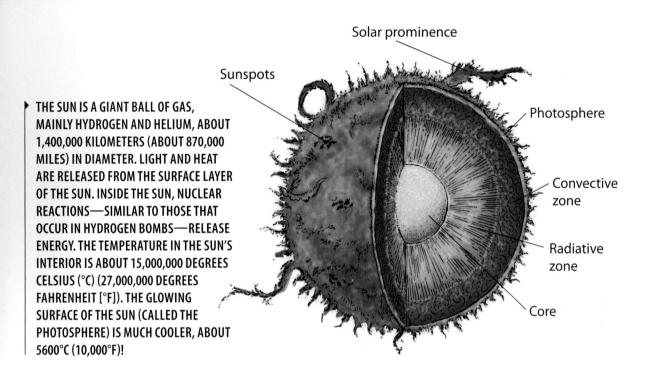

Solar prominence

Sunspots

Photosphere

Convective zone

Radiative zone

Core

▶ THE SUN IS A GIANT BALL OF GAS, MAINLY HYDROGEN AND HELIUM, ABOUT 1,400,000 KILOMETERS (ABOUT 870,000 MILES) IN DIAMETER. LIGHT AND HEAT ARE RELEASED FROM THE SURFACE LAYER OF THE SUN. INSIDE THE SUN, NUCLEAR REACTIONS—SIMILAR TO THOSE THAT OCCUR IN HYDROGEN BOMBS—RELEASE ENERGY. THE TEMPERATURE IN THE SUN'S INTERIOR IS ABOUT 15,000,000 DEGREES CELSIUS (°C) (27,000,000 DEGREES FAHRENHEIT [°F]). THE GLOWING SURFACE OF THE SUN (CALLED THE PHOTOSPHERE) IS MUCH COOLER, ABOUT 5600°C (10,000°F)!

▶ LIGHTNING IS PRODUCED WHEN STATIC ELECTRICITY IN CLOUDS DISCHARGES AND PRODUCES A SPARK. THIS SPARK IS SO HOT THAT IT SUPERHEATS THE AIR, CAUSING THE AIR TO EXPAND EXPLOSIVELY AND MAKE THUNDER.

PHOTO: kcdsTM/creativecommons.org

▶ IN GLOW STICKS, A CHEMICAL REACTION PRODUCES LIGHT, BUT NOT MUCH HEAT. WHEN THE STICK IS BENT, A GLASS VIAL INSIDE IT BREAKS. THIS ALLOWS THE REACTANTS TO MIX. THE PROCESS OF PRODUCING LIGHT BY THIS TYPE OF CHEMICAL REACTION IS CALLED CHEMOLUMINESCENCE.

PHOTO: Erik Charlton/creativecommons.org

FIRES, CANDLES, AND OIL LAMPS RELEASE LIGHT WHEN THE FUEL IN THEM REACTS WITH OXYGEN. HEAT AND LIGHT ARE PRODUCED IN THIS CHEMICAL REACTION.

PHOTO: ItzaFineDay/creativecommons.org

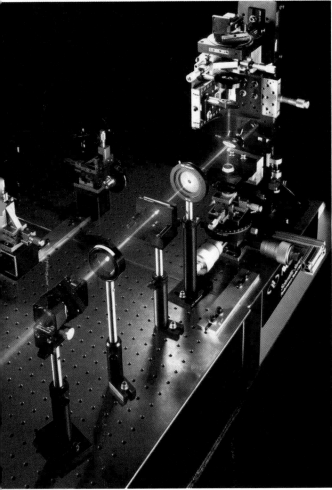

IN THIS LASER, LIGHT IS PRODUCED WHEN A FLASH OF LIGHT IS USED TO EXCITE ATOMS INSIDE A SPECIAL TUBE IN THE LASER. THE EXCITED ATOMS PRODUCE AN INTENSE BEAM OF SINGLE-COLORED LIGHT. LASER LIGHT HAS SPECIAL PROPERTIES THAT MAKE IT A VERY USEFUL LIGHT SOURCE.

PHOTO: Courtesy of DOE/NREL, Credit—Coherent Inc., Laser Group

DISCUSSION QUESTIONS

1. Make a list of all the devices in your home that emit light.

2. Which of the objects that you read about produce light and heat? Which produce light and little heat? Which type of object is a more efficient producer of light?

Life Light

Many kilometers down—in the deepest parts of the oceans—there is no sunlight. Down here in the darkness, thousands of species of organisms thrive, from single-celled microbes to giant worms and squid. Water covers much of the Earth. A huge volume of water makes up the oceans and seas, so it is not surprising that this is where you will find most of Earth's living things.

Most of this water lies in darkness. But some organisms living down here have eyes. Why have eyes if there is no light to see by? While no sunlight reaches these depths, there is light. Many marine organisms make their own light by a process called bioluminescence—"life light." They use this light to communicate with their own species or, in some cases, as a lure to capture their prey.

For example, the flashlight fish has glowing body parts! Under its eyes are pockets that contain bioluminescent microbes (bacteria). These bacteria make light all the time, but the fish hides these glowing pockets under movable flaps of skin. When the fish wants to reveal its lights, it simply moves the flaps.

But how do organisms make light? They don't use lightbulbs, but they do have something in common with flashlights. In a flashlight, part of the chemical energy in the batteries is transformed (via electrical energy) in the lightbulb to light and heat energy. Bioluminescent organisms also transform

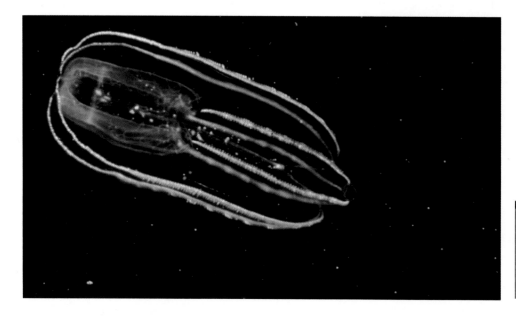

▶ **THIS LOBATE CTENOPHORE GIVES OFF A BIOLUMINESCENT GLOW.**

PHOTO: OAR/National Undersea Research Program (NURP)

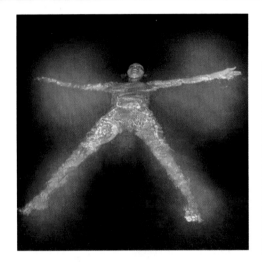

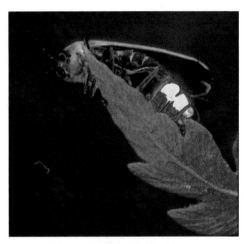

THE WATERS OF MOSQUITO BAY, PUERTO RICO, CONTAIN ABOUT 200,000 GLOWING MICROBES PER LITER. WHEN DISTURBED, AS BY THIS SWIMMER, THE MICROBES RELEASE A BLUE LIGHT. SADLY, THIS PERFORMANCE IS ENDANGERED BY POLLUTION.

PHOTO: Doug Myerscough

chemical energy into light. They use special enzymes—chemicals that speed up the chemical reaction—in this process. These enzymes cause a chemical called luciferin to react with oxygen. Light energy is released during this reaction.

Bioluminescence is very common in marine organisms. Many fish and squid glow. So do some microscopic organisms that float in the water. These are sometimes present in such large numbers that they can make a whole sea glow. For example, Mosquito Bay, Puerto Rico, is famous for its nightly performance of bioluminescence.

Organisms living in the depths of the oceans are not the only species that make their own light. Some nonmarine organisms also make light. Have you ever seen fireflies or lightning bugs? They use bioluminescence to communicate with each other. Try communicating with them on a warm summer's evening by using a flashlight. You may be surprised how many will reply to a short flash of light!

Go for a nighttime walk in the woods and you may observe rotting trees that glow in the dark. Glowing fungi inhabit the trees. Why do they glow? Perhaps to attract insects that can spread their spores.

Keep your eyes open. You may see other organisms glow. ■

FIREFLIES USE CHEMICAL REACTIONS WITHIN THEIR CELLS TO TRANSFORM CHEMICAL ENERGY FROM FOOD TO LIGHT ENERGY.

PHOTO: James E. Lloyd, Department of Entomology and Nematology, University of Florida, Gainesville

 DISCUSSION QUESTIONS

1. What functions does bioluminescence serve for organisms?

2. Research other bioluminescent organisms and how they use this characteristic.

HOW DOES LIGHT TRAVEL?

INTRODUCTION

In the previous lesson, you discussed light as being a type of energy. You discovered that light is made as a result of energy transformations. For example, in the lightbulb of the flashlight, electrical energy was transformed into light energy. Once light energy is produced, where does it go and how does it get there? In this lesson, you will observe light as it travels through and strikes different substances. You also will discuss how fast light travels.

THE SUN PROVIDES THE LIGHT THAT ILLUMINATES OUR PLANET. WHY DOES THE SPACE BETWEEN EARTH AND THE SUN LOOK SO DARK?

PHOTO: NASA

OBJECTIVES FOR THIS LESSON

Try to detect light as it travels through air and water.

Determine how light behaves as it travels.

Discuss how to determine the speed of light.

▶ **MATERIALS FOR LESSON 3**

For your group

1   copy of Student Sheet 3.1: Looking at How Light Travels

For your group

1   flashlight
2   D-cell batteries
1   plastic soda bottle
1   length of rubber tubing
1   dropping bottle

GETTING STARTED

1 One member of your group should collect the materials as directed by your teacher.

2 Remove the flashlight and shine it toward the wall. Answer the following questions in the Getting Started section of Student Sheet 3.1: Looking at How Light Travels:

A. Does light from the flashlight reach the wall?

B. How do you know?

C. Look at Figure 3.1. Where is light from the flashlight? Draw on the picture on Student Sheet 3.1 to show where you could find light from the flashlight.

D. Could you see light from the flashlight between the flashlight and the wall?

E. If not, how do you know it is there? (How could you detect it?)

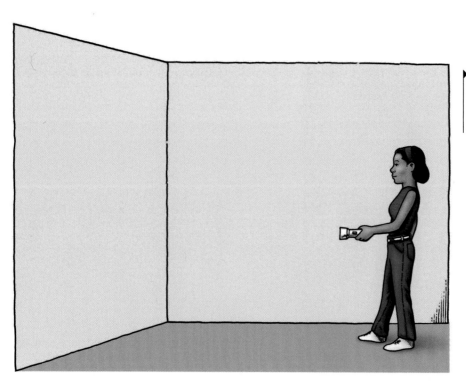

THIS FLASHLIGHT IS SWITCHED ON. WHERE CAN YOU FIND THE LIGHT IT IS PRODUCING?
FIGURE **3.1**

LOOKING AT HOW LIGHT TRAVELS

PROCEDURE

1 Use the plastic bottle and water to determine whether light travels through water. Answer the following questions on Student Sheet 3.1: Looking at How Light Travels:

A. How did you detect light on the side of the bottle opposite the flashlight?

B. Could you see light from the flashlight as it traveled through the water?

C. If you think light was passing through the water, but was invisible, how could you make it visible?

2 Test one of the suggestions from the class discussion.

A. Draw a labeled picture showing what you observed. Accurately describe the shape and the edges of the beam.

3 From what you have seen, do you think light travels in straight lines? Devise an experiment to test your hypothesis using the apparatus provided by your teacher. Conduct your experiment.

A. Use the outline in Table 1 on Student Sheet 3.1 to help you describe in words and diagrams the experiment you designed and what you discovered. Be prepared to share what you did and what you found out with the class.

REFLECTING
ON WHAT
YOU'VE DONE

1 Think about the points listed below. Discuss them with other students.

A. Does light travel through air and water?

B. What do you need to do to detect light?

C. Does light travel in straight lines? What evidence supports your conclusion?

2 Now consider this question: How fast does light travel? Discuss with your group how you could determine the speed of light. Here are some questions to think about:

A. What would you need to measure?

B. How would you make these measurements?

C. How accurate would your measurements need to be?

3 Read "Racing to Find the Speed of Light" on pages 36-39. Participate in a class discussion about what you read.

4 Write a paragraph in your science notebook summarizing what you have observed and discussed during this lesson. Be prepared to share your paragraph with the class.

RACING
to Find the Speed of Light

Light travels so fast that our eyes can't detect its movement. In the past, some people thought light traveled instantly from one place to another. Others thought light had a very fast, but finite speed—a specific speed that could somehow be measured. How do you measure such a fast speed? Think about it.

GALILEO'S GLIMMER

Galileo, the famous 17th-century scientist, asked just that question. And he had a plan to answer it. Galileo's plan involved two lanterns, a timing device, and a willing assistant. The assistant had to be willing—his job was to walk many kilometers and hike up a steep hill carrying one of the lanterns and the timing device.

Once on top of the hill, the assistant waited until dark. He covered the lantern so no light could be seen from it, and carefully lit it. Meanwhile, a nicely rested Galileo looked toward the hill and waited. He too had a lit, covered lantern. The assistant started the timer and uncovered his lantern. As soon as Galileo saw the light, he pulled the cover off his lantern. The assistant saw the distant glimmer. He tried to measure the time it had taken for the light to travel to Galileo and back to him. Unfortunately, the light had traveled too quickly for the timer to make an accurate measurement.

Galileo's method for determining the speed of light seemed like a good idea—but it was a failure. However, Galileo did learn something. He discovered that the speed of light was very fast. So fast that it could not be measured over such a short distance using such a simple timing device.

▸ WHO IS THIS STRANGER WITH THE HIDDEN LANTERN? WHAT WAS HIS ROLE IN TRYING TO DETERMINE THE SPEED OF LIGHT?

ROEMER AND THE ORBIT OF IO

If you can't make a better timer, why not increase the distance? That's a good idea, but a huge distance would be needed—a distance on an astronomical scale. Astronomer Olaus Roemer took up this challenge.

Roemer was studying Io, a moon that orbits Jupiter once every 1.76 days. He thought that he could use Io to calculate the speed of light. He knew that moons orbit planets at about the same speed regardless of the time of year. Why, he asked himself, did Io sometimes appear to be in the wrong position? At times, Io appeared to be ahead or behind its predicted position in orbit by up to 10 minutes. Roemer noticed that Io was ahead of schedule when Jupiter was closer to Earth, and behind schedule when Jupiter was farther away.

Could the speed of the light reflected from Io to Earth have something to do with this mystery? Roemer realized that the difference in schedule could be explained if light traveled at a specific, rather than at an infinite, speed. If light traveled at a specific speed, the farther the light had to travel, the longer it would take it to reach a destination. He calculated the change in distance when Io was farthest away from Earth and divided it by the change in time from Io's predicted position. Presto! Roemer had calculated the speed of light as 220,000,000 meters per second (721,784,777 feet per second).

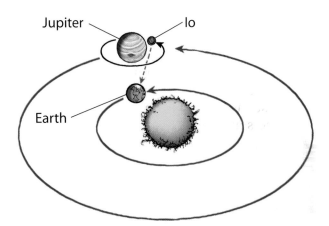

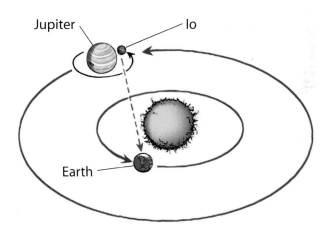

Note: Bodies and orbits are not to scale.

▶ THE FARTHER AWAY JUPITER AND ITS MOONS ARE FROM EARTH, THE LONGER IT TAKES LIGHT REFLECTED FROM THEM TO REACH US. ROEMER REALIZED THAT APPARENT VARIATIONS IN THE ORBIT TIME OF ONE OF JUPITER'S MOONS AND THE DIFFERENCE IN DISTANCE BETWEEN IT AND EARTH COULD BE USED TO CALCULATE THE SPEED OF LIGHT.

READING SELECTION

EXTENDING YOUR KNOWLEDGE

THE FINAL FIGURE

Roemer's calculation was pretty close, but scientists were determined to get a more accurate measurement. Many more attempts to determine the speed of light were made. The honor for the first truly accurate measurement went to the American physicist Albert Michelson. Like Galileo and Roemer, Michelson realized that to accurately measure the speed of light he needed to precisely measure the time it took light to travel a long distance. He used a specially designed apparatus with two mirrors that were 35 kilometers (21.7 miles) apart. Using a very accurate clock, he measured the speed of light at 299,799,600 meters per second (983,594,488 feet per second). Close, but not close enough for some scientists.

The most recent speed for light is 299,792,458 meters per second (983,571,056 feet per second). This measurement was made using very precise atomic clocks to calculate the time it takes for light to travel over very accurately measured distances. Science has come a long way since Galileo's experiment with lanterns and a timer! ■

▶ **HOW DID IO, A MOON ORBITING JUPITER, CONTRIBUTE TO OUR KNOWLEDGE OF THE SPEED OF LIGHT?**

PHOTO: NASA/Johns Hopkins Applied Physics Laboratory/Southwest Research Institute/ Goddard Space Flight Center

DISCUSSION QUESTIONS

1. How did each scientist's approach to measuring the speed of light produce a more accurate result than those who tried before him?

2. Why has measuring the speed of light been very challenging for scientists? What kinds of tools are needed to measure the speed of light?

LIGHT SPEEDSTERS

All these scientists attempted to find the speed of light. The accuracy of their results for the speed of light depended on their ability to design experiments and build apparatus that would accurately measure time and distance.

▶ GALILEO (1564–1642) TRIED UNSUCCESSFULLY TO MEASURE THE SPEED OF LIGHT WITH LANTERNS AND A TIMING DEVICE.

PHOTO: Library of Congress, Prints & Photographs Division, LC-USZ62-7923

▶ OLAUS ROEMER (1644–1710) USED ASTRONOMICAL DISTANCES TO CALCULATE THE SPEED OF LIGHT.

PHOTO: Library of Congress, Prints & Photographs Division, LC-USZ62-124161

▶ ARMAND FIZEAU (1819–1896) DESIGNED AN APPARATUS THAT USED ROTATING MIRRORS AND A DISTANCE OF 9 KILOMETERS (5.6 MILES) TO MAKE A MORE ACCURATE MEASUREMENT FOR THE SPEED OF LIGHT.

PHOTO: Courtesy of Smithsonian Institution Libraries, Dibner Library of the History of Science and Technology, Washington, D.C.

▶ ALBERT MICHELSON (1852–1931), A U.S. PHYSICIST AND NOBEL PRIZE WINNER, USED MIRRORS 35 KILOMETERS (21.7 MILES) APART TO OBTAIN WHAT IS CONSIDERED TO BE THE FIRST REALLY ACCURATE MEASUREMENT FOR THE SPEED OF LIGHT.

PHOTO: Courtesy of Smithsonian Institution Libraries, Dibner Library of the History of Science and Technology, Washington, D.C.

Light travels very fast. So fast, in fact, that to us it appears to travel instantly from one place to another. However, light does have a finite speed. Light travels at 299,792,458 meters (about 300,000 kilometers or 186,000 miles) per second in a vacuum. Light is slower when passing through transparent materials,

▶ COMPARED WITH OTHER GALAXIES, THE ANDROMEDA GALAXY IS CLOSE TO EARTH. LIGHT FROM THE ANDROMEDA GALAXY TAKES 2.2 MILLION YEARS TO REACH EARTH. ESTIMATES ARE THAT LIGHT FROM THE MOST DISTANT OBJECTS IN THE UNIVERSE TAKES OVER 13 BILLION YEARS TO REACH EARTH.

PHOTO: NASA/JPL/California Institute of Technology

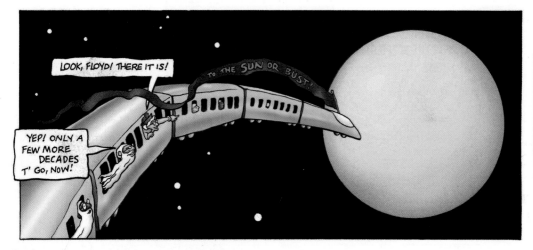

▶ A HIGH-SPEED TRAIN, TRAVELLING AT ABOUT 240 KILOMETERS (149 MILES) PER HOUR, WOULD TAKE ABOUT 75 YEARS— AN ENTIRE LIFETIME—TO TRAVEL TO THE SUN. YET, IT TAKES LIGHT ONLY 8.3 MINUTES TO COVER THE SAME DISTANCE.

such as air, water, or glass. Even so, it is difficult to imagine this kind of speed. To get an idea of how fast light travels, compare the time it takes for light to travel from different objects to the boy pictured in bed.

Because light is so fast, the distance it travels over a period of one year is sometimes used as a measure of astronomical distances. This unit of measure is called a light year. The light year unit is sometimes used by astronomers to measure the vast distances between stars and even between galaxies. A light year is therefore a measure of distance, not time! One light year is equal to about 9,500,000,000,000 kilometers (9.5×10^{12} km). ■

DISCUSSION QUESTIONS

1. What does a light year measure? Why do astronomers prefer to use light years rather than meters or kilometers?

2. Consider why you can never see anything in outer space exactly how it is at that moment. What about things on Earth?

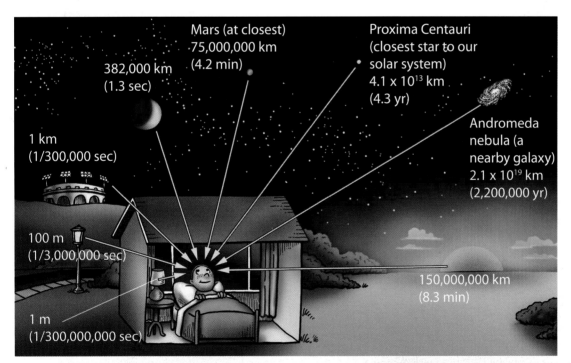

Mars (at closest)
75,000,000 km
(4.2 min)

Proxima Centauri
(closest star to our solar system)
4.1×10^{13} km
(4.3 yr)

382,000 km
(1.3 sec)

Andromeda nebula (a nearby galaxy)
2.1×10^{19} km
(2,200,000 yr)

1 km
(1/300,000 sec)

100 m
(1/3,000,000 sec)

150,000,000 km
(8.3 min)

1 m
(1/300,000,000 sec)

▶ WHEN YOU LOOK AT THE MOON, YOU ARE ACTUALLY SEEING IT AS IT WAS 1.3 SECONDS AGO. IF YOU GLANCE UP AT THE SUN, YOU ARE SEEING IT AS IT WAS ABOUT 8.3 MINUTES AGO. THAT'S HOW LONG IT TOOK THE LIGHT THE BOY SEES TO TRAVEL TO THE BOY'S EYES. IF YOU USED A TELESCOPE TO VIEW DISTANT GALAXIES, YOU WOULD SEE THEM AS THEY WERE MILLIONS OF YEARS AGO! THE FINITE SPEED OF LIGHT ALLOWS ASTRONOMERS TO SEE INTO THE PAST AND UNRAVEL THE HISTORY OF OUR UNIVERSE.

Light and Distance

Why do lights look dimmer when they are farther away? To answer this question we need to think about how light spreads out from a light source. First think about what happens when you quickly turn a lightbulb on and off. A flash, or pulse, of light is produced. This pulse of light can be thought of as moving away from the lightbulb in all directions, like a shell or hollow sphere of light. The farther the pulse of light is from the lightbulb, the bigger the hollow sphere of light becomes.

As the sphere of light gets bigger, the light spreads out. The area illuminated by the pulse of light increases very quickly. The amount of light reaching each square centimeter gets smaller. This explains why the farther you get from a light source, the dimmer it appears.

The apparent brightness of light, or how it looks to your eyes, decreases much more quickly than you might think. This is because the area over which the light spreads—the surface area of the hollow sphere of light—grows faster than the distance the light travels. When you double the distance from a light source such as a lightbulb (say from 10 to 20 cm or 3.9 to 7.9 inches), the light spreads out over an area four times as large. Observed from this distance, the lightbulb appears one-fourth as bright. Double the distance again, this time to 40 cm (15.7 inches), and the same lightbulb appears only one-sixteenth as bright!

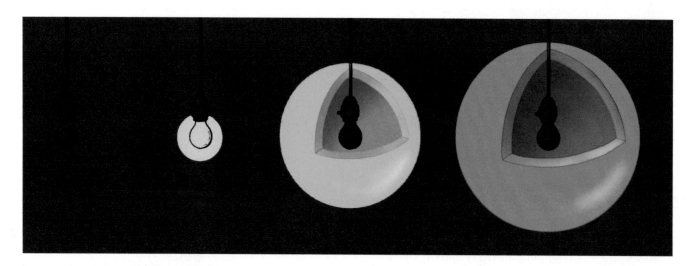

▶ **WHEN THIS LIGHTBULB FLASHES, LIGHT SPREADS OUT FROM IT IN ALL DIRECTIONS LIKE A SHELL OR HOLLOW SPHERE OF LIGHT.**

This rapid change in the apparent brightness of light as you move away from its source is sometimes called the inverse square law, which is shown in the following equation:

APPARENT BRIGHTNESS = BRIGHTNESS AT SOURCE/DISTANCE²

This law explains why bright street lamps can be seen for only a few miles, and why it gets dark quickly when you walk away from a nighttime campfire. ■

DOING THE MATH ON SPREADING LIGHT

The following equations give some mathematical proof for inverse square law:

The area illuminated by a sphere of light is
4 × π × radius of the sphere²

If the approximate value of π is
taken as 3.14, then:
The area illuminated by a sphere of light with a radius
of 10 cm is
4 × 3.14 × 10 × 10 = 1256 cm²

The area illuminated by a sphere of light
with a radius of 20 cm is
4 × 3.14 × 20 × 20 = 5024 cm²

What will be the area illuminated by a sphere of light
with a radius of 30 cm?

DISCUSSION QUESTIONS

1. How is the behavior of a pulse of light from a lightbulb like inflating a balloon?

2. We have learned that light becomes rapidly dimmer as we travel farther from its source. How do our perceptions of sound and odor change as we increase our distance from their respective sources?

BLOCKING THE LIGHT

INTRODUCTION

Light can pass through the vacuum of space. You know this because light from the Sun reaches Earth. You have observed during your experiments that light can travel through air and water, some types of plastic, and glass. But can light travel through all types of matter? Does it pass through different types of matter in the same way? In this lesson, you will try to answer these questions.

▶ WHERE'S THE SHADOW OF THE TURTLE? HOW DO YOU THINK SHADOWS ARE FORMED? IS THE SHADOW OF SOMETHING ALWAYS THE SAME SIZE AND SHAPE? CAN SOMETHING HAVE MORE THAN ONE SHADOW?

PHOTO: Jason Pratt/ creativecommons.org

OBJECTIVES FOR THIS LESSON

Observe what happens when light strikes different materials.

Investigate and explain the appearance and formation of shadows.

▶ MATERIALS FOR LESSON 4

For you

1	copy of Student Sheet 4.1: Putting Objects in the Path of Light
1	copy of Student Sheet 4.2: Measuring Shadow Size

For your group

1	assembled light stand
1	meterstick
1	white screen
2	plastic stands
1	black disk
1	craft stick
1	bag containing samples of paper, plastic, cardboard, and other objects
1	transparency
1	marker
	Masking tape

GETTING STARTED

1. Have one member of your group obtain the bag of objects.

2. Spread the contents of the bag out on a desk.

3. Examine the objects. Work with your group to sort the objects into groups according to how they allow light to pass through them.

4. Create a table in your science notebook to list your groups and record the objects in each. 📝

5. Discuss your groupings with the class.

6. Work with the class to develop a definition for the different groups. After the class has decided on a definition for each group, record those definitions.

INQUIRY 4.1

PUTTING OBJECTS IN THE PATH OF LIGHT

PROCEDURE

1. Have one member of your group obtain the materials as directed by your teacher.

2. Record your observations, measurements, and explanations on Student Sheet 4.1: Putting Objects in the Path of Light.

3. Working with your group, set up the screen. Place the light stand so the lightbulb is horizontal. Position the lightbulb so the tip of its filament is 50 cm away from the screen. Put your hand between the lightbulb and the screen (see Figure 4.1).

SAFETY TIP

Lightbulbs become hot very quickly and take many minutes to cool down. Make sure your lightbulb is cool before you handle it.

▶ **WHICH PARTS OF THIS FROG CAN LIGHT PASS THROUGH?**

PHOTO: By Jurvetson (flickr)

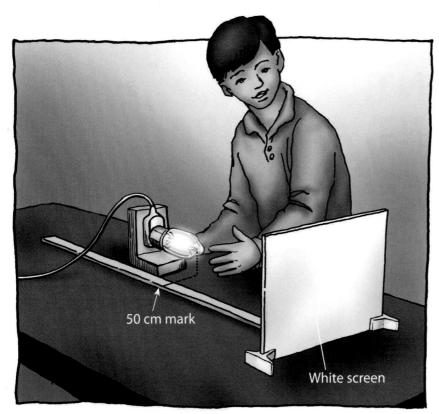

50 cm mark

White screen

▶ SET UP THE SCREEN. PLACE THE LIGHTBULB
IN A HORIZONTAL POSITION. PUT YOUR HAND
BETWEEN THE SCREEN AND THE LIGHTBULB.
FIGURE **4.1**

4 Switch on the lightbulb. Record what you observe on the screen. You may also draw a diagram to explain what you see.

5 Hold each object from "Getting Started" about 20 cm from the screen. Create a table, and record your descriptions of each object.

6 Discuss your observations with your group, and write an explanation of what you observe. Record what you think causes shadows to form. Which of your objects produced the darkest shadows?

READING SELECTION

BUILDING YOUR UNDERSTANDING

TRANSPARENT, TRANSLUCENT, AND OPAQUE

Substances that let light of certain wavelengths pass through them and can be clearly seen through are called transparent. Water, glass, and acetate sheets are transparent. Colored glass is also transparent—it allows some colors through but not others. Materials that let light pass through but cannot be clearly seen through are called translucent. These materials are said to diffuse, or scatter, light. Wax paper and frosted glass are two examples of translucent substances. Materials that do not let light pass through them are called opaque. Sometimes, thin samples of opaque materials (thin paper, for example) may be translucent in bright light.

▶ WATER IS CONSIDERED TRANSPARENT, BUT IT DOES ABSORB AND SCATTER SOME LIGHT. THIS IS ONE REASON WHY LIGHT DOES NOT REACH THE DEEPEST PARTS OF THE OCEANS. SUBMARINERS MUST CARRY LIGHT WITH THEM.

PHOTO: OAR/National Undersea Research Program (NURP); Harbor Branch Oceanographic Institute

▶ CLOUDS MAY ALLOW LIGHT TO PASS THROUGH THEM, BUT YOU CANNOT SEE THROUGH THEM. THEY ARE TRANSLUCENT.

PHOTO: NOAA Photo Library, NOAA Central Library; OAR/ERL/National Severe Storms Laboratory (NSSL)

All matter interacts in some way with light. Water, for example, is fairly transparent, but it does absorb and scatter some light. As a diver goes deeper into the sea, the color of the light from the Sun changes—it gets bluer and darker. Sunlight cannot penetrate the deepest parts of the oceans. ■

MEASURING SHADOW SIZE

PROCEDURE

1 Record all your observations, measurements, and explanations on Student Sheet 4.2: Measuring Shadow Size.

2 Switch on the lightbulb. Put your hand between the screen and the lightbulb, and move your hand backward and forward between the lightbulb and the screen. Write a description of what you observe.

3 Measure and record the diameter of the black disk. Attach the black disk to the craft stick (see Figure 4.2).

4 Use the disk to make a shadow on the screen. Hold the disk 20 cm from the screen, and draw the shadow produced.

5 You are going to measure the size of the shadow made by the disk when it is placed at different distances from the screen. Start by measuring the diameter of the shadow produced when the disk is 20 cm from the screen. You should make at least five measurements of shadow diameter with the disk at different distances from the screen. Design and draw a table for your results.

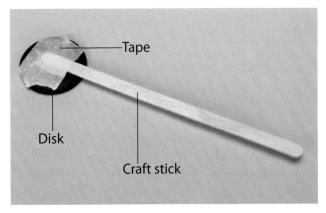

▶ **USE TAPE TO ATTACH THE DISK TO THE CRAFT STICK.**
FIGURE **4.2**
PHOTO: ©2009 Carolina Biological Supply Company

Inquiry 4.2 continued

6 Choose five positions at different distances from the screen. Measure the diameter of the shadow produced at each position (see Figure 4.3). Record your results in the table.

7 Present your results as a graph using the grid provided on Student Sheet 4.2.

8 Is there a relationship between the diameter of the shadow produced by the disk and the distance of the disk from the screen? Record your ideas.

9 Draw a diagram to show what you observed. Be prepared to use your diagram to share your findings with the class.

50 cm mark

▶ PLACE THE DISK AT DIFFERENT DISTANCES FROM THE SCREEN. MEASURE THE DIAMETER OF THE SHADOW THE DISK PRODUCES.
FIGURE **4.3**

INQUIRY 4.3

COMPARING SHADOWS

PROCEDURE

1 Place the lightbulb in a vertical position, as shown in Figure 4.4. Use the disk to make a shadow on the screen. Record your responses for this inquiry in your science notebook.

A. Draw a diagram of the shadow produced.

B. Describe how the shadow formed when the lightbulb is placed vertically differs from the shadow formed when the lightbulb is placed horizontally.

C. Have you ever seen a shadow like this before? If so, where?

2 After a class discussion about shadows, add labels to your diagram from Step 1A.

3 Draw a diagram that explains how fuzzy shadows are produced.

REFLECTING
ON WHAT
YOU'VE DONE

1 After discussing the questions below with your group, record your responses in your science notebook:

A. How are shadows produced?

B. What is the relationship between the size of a shadow on the screen and the distance of the screen from the object that produces the shadow?

C. What is the relationship between the size of a shadow on the screen and the distance of the object from the light source?

D. Which types of light sources produce the sharpest shadows?

2 Is it possible for an object to cast more than one shadow? Share materials with other pairs and groups to help you answer this question.

A. Write a short paragraph about what you discovered.

3 Review the question bank created in Lesson 1. Can you answer any more questions now? Identify those that you feel comfortable answering.

50 cm mark

▶ MOVE THE LIGHT STAND SO THAT THE LIGHTBULB IS IN A VERTICAL POSITION. USE THE DISK TO MAKE A SHADOW.
FIGURE **4.4**

ASTRONOMICAL SHADOWS

Shadows are areas of darkness that form behind an object when the object blocks a source of light. When shadows are formed on an astronomical scale, they can have a dramatic effect. They can produce eclipses of the Sun and Moon that we can see on Earth.

From Earth, two types of eclipses are visible to the naked eye—lunar eclipses and solar eclipses. A lunar eclipse occurs when Earth comes between the Sun and Moon and casts a shadow on the Moon. A solar eclipse happens when the Moon casts its shadow on Earth. ■

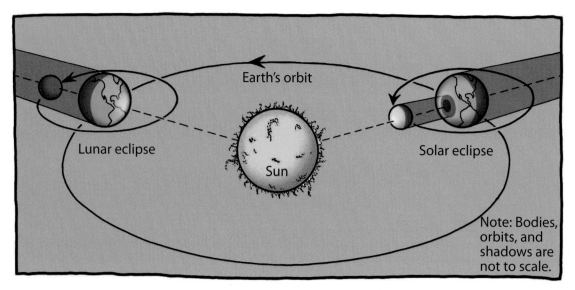

Earth's orbit

Lunar eclipse

Sun

Solar eclipse

Note: Bodies, orbits, and shadows are not to scale.

▶ THE SHADOWS MADE BY AN ECLIPSE HAVE TWO REGIONS, THE UMBRA AND PENUMBRA. YOU MAY HAVE OBSERVED THESE REGIONS OF A SHADOW IN INQUIRY 4.3. DURING A SOLAR ECLIPSE, THE AREA UNDER THE UMBRA IS IN ALMOST COMPLETE DARKNESS. THE AREA UNDER THE PENUMBRA BECOMES ABOUT AS DARK AS A DARK, CLOUDY DAY.

▶ A LUNAR ECLIPSE IN PROGRESS: A SERIES OF PICTURES TAKEN AT DIFFERENT TIMES DURING AN ECLIPSE. YOU CAN SEE THE SHADOW OF EARTH MOVING ACROSS THE MOON'S SURFACE. WHY IN A TOTAL ECLIPSE DOES THE MOON LOOK RED? A LATER LESSON WILL HELP YOU ANSWER THIS QUESTION.

PHOTO: NASA

THEATER OF SHADOWS

It began nearly one thousand years ago on the island nation of Indonesia. Shadow masters, called *dalangs*, traveled from village to village entertaining people with tales of love and war.

They used flat puppets, made of water buffalo hides, to play the parts of brave princes, evil brutes, and willful heroines. The shadows of the dancing and fighting puppets were created by placing the puppets between the light of an oil-burning lamp and a thin cloth screen.

The villagers looked forward to the *dalang*'s visits. They gathered at the village square or temple yard at sunset to watch and listen until dawn of the next day! To the people of Indonesia, the long plays were well worth it. The stories were morality tales that reminded the adults, and taught the children, about the nature of good and evil. And they believed that the puppets' shadows were their ancestors' spirits returning to Earth.

THE MAGIC AND THE STORIES

Tamara Fielding was born in Indonesia on the island of Java. As a child, she was not supposed to go behind the *dalang*'s screen to watch him perform. Only men and boys were allowed to do that. But young Tamara couldn't resist. She secretly crept behind the screen and watched the *dalang* work. His skillful and graceful performance amazed her. So did the beauty and fierceness of the puppets. From that early exposure, the "magic and the stories of shadow theater were locked inside me," says Tamara. And there they would stay for many years.

Tamara moved to Europe in her 20s. She studied drama and theater in Paris. Eventually, she came to the United States, where she continued her acting career. More years went by. Then one day, Tamara decided to dust off the shadow puppets that relatives had given her years before. She built a cloth screen and started moving the puppets in a way she remembered the *dalang* doing so long ago. The magic and the stories inside her began to come out.

▶ AS A CHILD IN INDONESIA, TAMARA WAS FASCINATED BY SHADOW PUPPET THEATER. NOW SHE SHARES HER ENJOYMENT AND KNOWLEDGE OF IT WITH A NEW GENERATION OF YOUNGSTERS.

PHOTO: Photos of Javanese shadow puppets from the private collection of Tamara Fielding and Tamara and the Shadow Theatre of Java; www.indonesianshadowplay.com

SHADOW FIGURES

Now Tamara has more than 400 puppets, or "shadow figures," as she calls them. She travels—not from village to village—but to universities, museums, and festivals sharing the stories and art of shadow theater.

Wearing a sarong and lace blouse with a flower in her hair, Tamara sits on the floor between the screen and the light. Unlike *dalangs* of old, she uses a 600-watt halogen lamp to create her shadows. And most of her performances last less than an hour—not all night.

▶ THIS TRADITIONAL INDONESIAN PLAY IS ABOUT ANCESTORS' SPIRITS RETURNING TO EARTH AT NIGHT. THE COLORS OF THESE PUPPETS ARE NOT VISIBLE. THE AUDIENCE SEES ONLY THE PUPPETS' SHADOWS.

PHOTO: Photos of Javanese shadow puppets from the private collection of Tamara Fielding and Tamara and the Shadow Theatre of Java; www.indonesianshadowplay.com

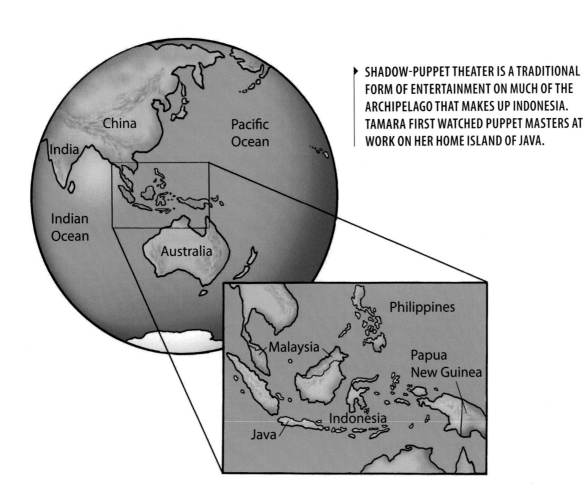

SHADOW-PUPPET THEATER IS A TRADITIONAL FORM OF ENTERTAINMENT ON MUCH OF THE ARCHIPELAGO THAT MAKES UP INDONESIA. TAMARA FIRST WATCHED PUPPET MASTERS AT WORK ON HER HOME ISLAND OF JAVA.

Today the shadow figures are as lively and powerful as ever. Like *dalangs* of old, Tamara creates the many voices of her characters. Her varied cast includes heroes, princesses, and animals, including monkeys, tigers, elephants, birds, and snakes.

With a single light source and a piece of cloth, the many dark shapes come alive. In the magical world of shadow theater, the shadows—magnificent, beautiful, and loathsome—take center stage. ■

DISCUSSION QUESTIONS

1. How are Tamara's shadow-puppet shows different from those of her ancestors? How are they similar?

2. Besides shadows, what other lighting techniques are used to enhance performance art?

WHERE DOES COLOR COME FROM?

INTRODUCTION

In the fall, hundreds of thousands of tourists from all over the world go to New England to enjoy the changing colors of the leaves. Color is very important to them, but then, it is important to everyone. Whether you are shopping for new clothes or a car, following a marked trail in the woods, or waiting for a red light to turn green, color matters. Color provides you with a great deal of useful information about your environment. The universe would certainly be a dull place without it. So, what is color? Why do things appear to be colored? Where does color come from? When you see color, what exactly are you seeing? In this lesson and several that follow, you will try to answer some of these questions as you investigate color.

▶ **WHEN PEOPLE ENJOY THE FALL COLORS, WHAT ARE THEIR EYES DETECTING?**

PHOTO: Ctd 2005/creativecommons.org

OBJECTIVES FOR THIS LESSON

Investigate where color comes from.

Discuss the appearance of the visible spectrum.

▶ **MATERIALS FOR LESSON 5**

For your group and another group to share

1	ray box
1	ray box lid
1	60-W clear halogen lightbulb
1	extension cord
1	bulb holder
2	narrow-slit ray box masks
2	no-slit ray box masks

For your group

1	triangular prism
1	flashlight
2	D-cell batteries
1	white screen
2	plastic stands
1	box of colored pencils

GETTING STARTED

1 Collect your materials as directed by your teacher. Take out the flashlight and triangular prism.

2 Have members of your group take turns holding the triangular prism close to their eye and looking around the room. Look through the window and at the lights in the classroom. Discuss with your partners where you have seen this effect before.

3 After your teacher has darkened the room, work with your group to shine the flashlight through the prism onto the ceiling. Try holding the prism at different distances from the flashlight and turning the prism around in the beam of light. What do you observe?

4 Discuss your observations with your group and with the class.

▶ **WHAT DOES A TRIANGULAR PRISM DO TO LIGHT?**

PHOTO: Nancy Rodger, © Exploratorium, www.exploratorium.edu

INQUIRY 5.1

USING A TRIANGULAR PRISM TO EXAMINE WHITE LIGHT

PROCEDURE

1 Record your responses for this inquiry in your science notebook. There is no student sheet for this lesson. 🖉

2 Watch as your teacher demonstrates how to set up the ray box, prism, and white screen. Follow the instructions set out in the diagrams labeled A–D in Figure 5.1 to construct your ray box.

SAFETY TIP

The lamp in the ray box gets hot. Allow it to cool for at least 5 minutes before handling the lightbulb.

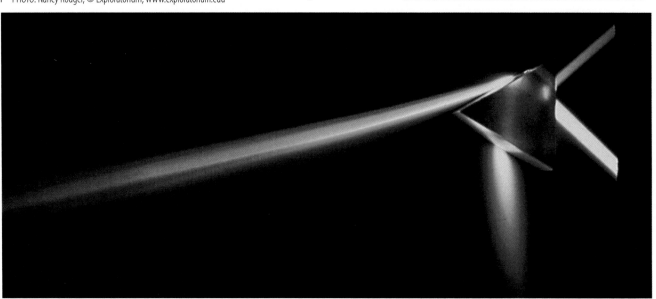

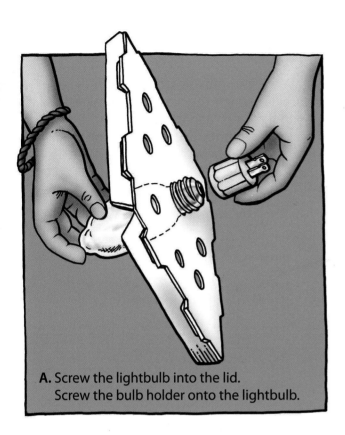

A. Screw the lightbulb into the lid. Screw the bulb holder onto the lightbulb.

B. Open the box. Put it on the table. Place the lid on the box. The flaps fit inside. Attach the extension cord to the bulb holder.

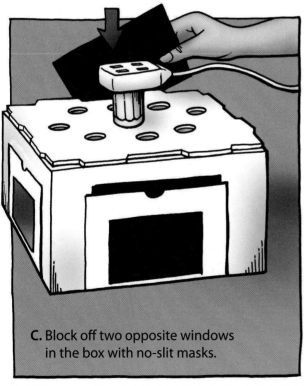

C. Block off two opposite windows in the box with no-slit masks.

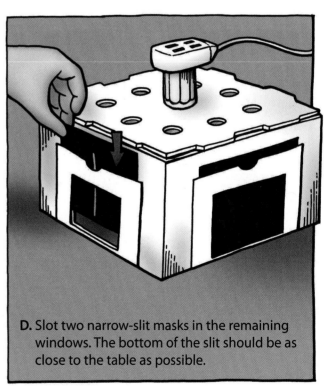

D. Slot two narrow-slit masks in the remaining windows. The bottom of the slit should be as close to the table as possible.

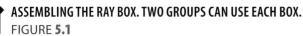

▶ ASSEMBLING THE RAY BOX. TWO GROUPS CAN USE EACH BOX.
FIGURE **5.1**

Inquiry 5.1 continued

3 Place the triangular prism in the path of the light ray. Draw what happens to the path of the light ray as it enters and leaves the prism.

4 Attach the plastic stands to the white screen. Place the screen in the path of the light ray leaving the prism. Rotate the prism and, if necessary, move the screen (as shown in Figure 5.2) until you get the same colors you observed in "Getting Started."

A. Use colored pencils to draw what you see.

B. Write down the names of the colors in the order in which you observe them.

C. Can you get the colors to appear on the screen in a different sequence?

Light ray

▶ CAN YOU GET THE SAME COLORS YOU OBSERVED IN "GETTING STARTED" TO FORM ON YOUR SCREEN?
FIGURE **5.2**

5 Where do you think the colors you observed came from? Discuss your ideas with your partners. Here are some questions to consider:

A. What happened to the light rays when they first entered the prism?

B. Did all the colors leave the prism at the same place?

C. What did the prism do to the white light?

SAFETY TIP

Turn off the ray box immediately after you finish using it. Allow the ray box to cool.

6 Write a paragraph explaining where you think the colors came from.

7 Review your observations and explanations with the class. You may be asked to read your paragraph aloud.

REFLECTING
ON WHAT
YOU'VE DONE

1 Scientists once thought that glass added colors to white light. What do you think of this hypothesis? How could you test it? Discuss this idea with your group. Be prepared to engage in a class discussion.

2 Read "The Impurity of White" on pages 62-63, and record your responses to the questions below in your science notebook.

A. What did Newton's experiments prove?

B. Write a paragraph summarizing his experiment.

The Impurity of White

For centuries the Western world had thought of white as the color of purity. The color white was associated with all that is pure and innocent. Along came a young English scientist, Isaac Newton (1642–1727), who changed people's ideas about the world of color forever. It was Newton who discovered in the 17th century that white light is not pure; it is a mixture of different colors.

Working in a darkened room, Newton experimented with a beam of light by allowing sunlight to shine into the room through a small hole in the wall. He then passed this beam of light through a glass prism. The result was a line of different colors, which he projected onto a white screen. But this part of his experiment was not new.

Many people had already observed that white light can produce a line of colors that is called a spectrum. For thousands of years, people observed how colors were produced when sunlight shone through crystals or droplets of water.

Everyone recognized these colors were the same as those found in rainbows. Most people believed that pure white light from the Sun was changed or contaminated in some way when it passed through the crystals or water. But on that day in 1666, Newton took his experiment one step further. And what he discovered turned people's ideas about the world of color upside down.

RECOMBINING THE SPECTRUM

"What would happen," Newton wondered, "if I used another prism but turned it upside down with respect to the first? Then the colors produced by the first prism would shine into the second one." He tried out his idea. He used two prisms and a lens, as shown here.

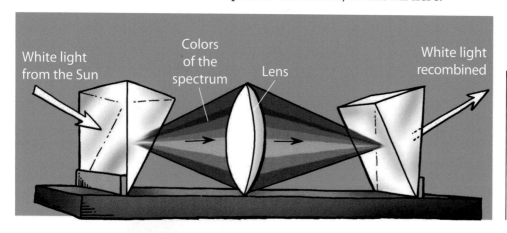

White light from the Sun

Colors of the spectrum

Lens

White light recombined

▶ NEWTON USED PRISMS AND A LENS TO RECOMBINE THE COLORS IN WHITE LIGHT. SUNLIGHT CAME IN FROM THE LEFT, WAS SPLIT INTO ITS SPECTRUM, AND THEN WAS RECOMBINED USING A LENS AND A SECOND, IDENTICAL, UPSIDE-DOWN PRISM.

DISCUSSION QUESTIONS

1. Why do you think people initially opposed Newton's discoveries about light?

2. Research other scientific discoveries that were initially rejected by a skeptical public. Why do you think those ideas were rejected?

▶ ALL COLOR COMES FROM LIGHT, WHETHER IT IS THE COLORED BEAK OF A SPECTACULAR BIRD LIKE THIS PENGUIN OR THE BRIGHT COLORS OF CLOTHES. ALL THINGS THAT ARE COLORED REFLECT SOME OF THE COLORS THAT MAKE UP WHITE LIGHT.

PHOTO (above, left): Courtesy of Carolina Biological Supply Company
PHOTO (left): © M.J.F. Marsland 2000

The first prism produced a colored spectrum. The lens then focused the spectrum onto a second, identical, upside-down prism. The spectrum left the second prism as a ray of white light. Newton could think of only one logical explanation. Instead of contaminating white light, the first prism split it into the different colors that make up white light. The second prism recombined those colors back into white light. White was not pure, but a mixture of color.

At the time many people found Newton's discovery very disturbing. It took quite a few years before they could accept that white was not a pure color. In fact, it was the least pure of all colors. At first, there was much opposition to his discovery. However, Newton's experiment was easily repeated. By the end of the 17th century, it was widely accepted that Newton's explanation of the spectrum produced by white light was the correct one. ■

Explaining a Rainbow

▶ PHOTO: Rhett Maxwell/creativecommons.org

White light is made up of the many colors of visible light—light we can see. A prism splits white light into these colors. It does this by bending the various colors of light at slightly different angles as they enter the prism. This causes the colors to spread out, splitting them apart.

You have seen these colors before. They are the same ones that make up a rainbow. But a rainbow does not contain a prism, so how are the colors of a rainbow made?

A rainbow is not made from a single prism like the one you used, but a summer shower does contain billions of raindrops. As sunlight enters each drop, it bends, bounces off the back of the raindrop, and bends again when it leaves. As you can see from the illustration, bending the sunlight makes the white light split—just like a prism bends and splits white light. The raindrops that produce a rainbow behave a bit like billions of tiny prisms.

But why does a rainbow look like big bands of color? Inside each raindrop, each color of white light is bent a different amount. Because the raindrops are in different positions in the sky, they bend different colors to your eye. The appearance of the rainbow depends on the raindrops that make it up. Larger droplets separate the colors well, making the colors of the rainbow appear more distinct. When the droplets are very small, the colors can overlap and the rainbow becomes almost white and more difficult to see.

Colored bows can be seen wherever sunlight strikes water droplets at certain angles. Many waterfalls produce rainbows in their spray. Sometimes bows of color can be seen on early morning dew. These are called "dew bows." ■

WHICH COLORS ARE IN A RAINBOW?

A rainbow has many colors. How many exactly? Some people suggest there are only seven: red, orange, yellow, green, blue, indigo, and violet. They even have a nonsense phrase to help them remember them—ROY G BIV. How many colors did you detect using your prism? Were there only seven? Could you see a greenish blue or a yellowish green? In fact, there are millions of colors in a rainbow. It's just that we have names for only a few of them!

Can you see all the colors our eyes can perceive in a rainbow or through a prism? Try looking for magenta or brown.

▶ YOU DON'T NEED RAIN TO MAKE A RAINBOW. THE SPRAY FROM A WATERFALL CAN PRODUCE A RAINBOW.

PHOTO: Marco Franchino/creativecommons.org

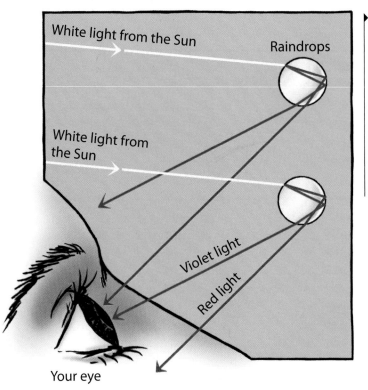

White light from the Sun

Raindrops

White light from the Sun

Violet light

Red light

Your eye

▶ RAINDROPS SPLIT WHITE LIGHT INTO THE COLORS THAT MAKE IT UP. RAINDROPS IN DIFFERENT POSITIONS SEND DIFFERENT COLORS TO THE EYE OF SOMEONE LOOKING AT A RAINBOW. RED LIGHT IS SEEN FROM RAINDROPS AT AN ANGLE OF 42° TO THE HORIZON, VIOLET LIGHT FROM RAINDROPS AT AN ANGLE OF 40°. THE OTHER COLORS OF THE RAINBOW ARE SEEN AT ANGLES BETWEEN 40° AND 42°. THIS IS WHY RAINBOWS ARE ONLY VISIBLE TO THE OBSERVER AS A NARROW CURVED BAND IN THE SKY. USING THIS DIAGRAM, THINK ABOUT WHAT DIRECTION—UP OR DOWN—YOU WOULD HAVE TO LOOK TO SEE RED LIGHT OR VIOLET LIGHT.

DISCUSSION QUESTIONS

1. Discuss where you have to be relative to the Sun and rain to see a rainbow.

2. Discuss how you would use five prisms to create a device to shine a rainbow on a wall.

COLOR, WAVELENGTH, AND THE WIDER ELECTROMAGNETIC SPECTRUM

INTRODUCTION

In the previous lesson, you discovered that white light could be split to produce a visible spectrum of different colors. What is it about light that makes it appear to be different colors? What property of light allows your eyes to distinguish between red and blue or between yellow and orange? In this lesson you will use the wave model of light to help explain what your eyes detect when you see color. You will relate the waves to the colors visible in the spectrum.

▶ **WHAT PROPERTY OF LIGHT ENABLES US TO DISTINGUISH BETWEEN THE COLORS ON THESE CARDS?**

PHOTO: Jeff McAdams, Photographer, Courtesy of Carolina Biological Supply Company

OBJECTIVES FOR THIS LESSON

Model light waves of different colors.

Construct a model of the visible electromagnetic spectrum.

Read about infrared and ultraviolet light.

▶ MATERIALS FOR LESSON 6

For you

1   copy of Student Sheet 6.2: Constructing the Spectrum

1   pair of scissors

For your group

1   piece of beaded chain

1   meterstick

     Glue

GETTING STARTED

1. Can you use a piece of beaded chain to make waves? As you and your partners experiment with the chain, think about these questions:

   A. What do you have to do to the chain to make waves?

   B. What does the wave transmit down the chain?

2. In a class discussion, share your approach to making waves and any observations you may have.

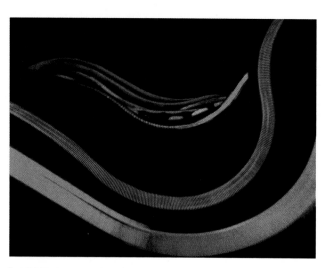

▶ **LIGHT IS A WAVE.**

PHOTO: Image by Kevin Dooley under Creative Commons license {http://www.flickr.com/photos/pagedooley/2415438447}

INQUIRY 6.1

MEASURING DIFFERENT WAVELENGTHS

PROCEDURE

1. You are going to examine and measure the waves you make with your chain. To do this, you will need to freeze the motion of the chain on a flat surface. Work with your classmates to determine which method of making waves best allows you to freeze the motion of the chain.

2. Use the method the class selects to make waves. Stop shaking the chain abruptly. Do not move your chain. In your science notebook, draw the pattern formed by the chain. ☞

3. Read "Measuring a Wave." Measure the wavelength (in mm) of the wave shown in the diagram in the reading selection. Record the measurement you take.

4. Now measure and record the wavelength of the wave you made with the chain (see Figure 6.1). Your wave may not be the exact shape of the wave shown in the diagram. You may find it useful to measure a few wavelengths on your chain and calculate the average (mean) wavelength.

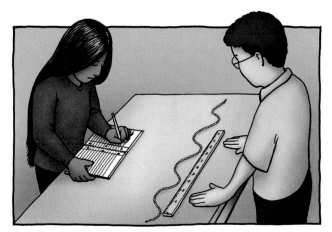

▶ WHAT IS THE WAVELENGTH OF THE WAVE YOU MADE WITH THE CHAIN?
FIGURE **6.1**

5 Shake the chain faster than you did in Step 2. Abruptly stop shaking it. Do not move the chain. Draw the pattern formed by the chain. Write answers to the following questions:

A. What is the difference between the two wave patterns?

B. What is the wavelength of the new wave?

C. What happened to the wavelength of the chain when you shook the chain faster?

6 Shake the chain again, starting slowly and then speeding up. Record what happens to the frequency of waves in the chain. (Remember: Frequency is the number of waves that pass a certain point every second.)

A. What is the relationship between the wavelength of your waves and their frequency?

B. Which wavelengths of your chain carry the most energy?

MEASURING A WAVE

Waves can vary in length and height. The distance between the crests (or the troughs) of two adjacent waves is the wavelength of the wave. The distance from the midpoint of the wave to its crest is the amplitude of the wave.

The number of waves that pass a certain point during a specific time—usually one second—is called the frequency of the wave, and is measured in Hertz (Hz). With light rays, the speed of the light—the rate at which the waves move along—depends on what substance (or material) the light waves are traveling through. Light travels slightly slower in air than in a vacuum, and its exact speed through air depends on the density of the air. However, because light travels at a constant speed through a vacuum or any one material, frequency and wavelength are related to one another. Remember, the shorter the wavelength, the higher the frequency. ■

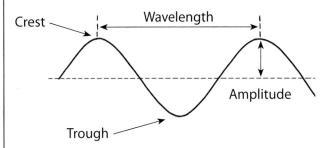

▶ THE DISTANCE BETWEEN TWO CRESTS, OR TWO TROUGHS, OF ADJACENT WAVES IS CALLED THE WAVELENGTH. THE DISTANCE FROM THE MIDPOINT OF THE WAVE TO ITS CREST IS THE AMPLITUDE OF THE WAVE.

INQUIRY 6.2

CONSTRUCTING THE SPECTRUM

PROCEDURE

1 Read "Color and Wavelength."

2 In this inquiry you will construct a model of the visible spectrum by identifying some waves and drawing other waves in the order they appear in the visible spectrum. Remember that red light has the longest wavelength and violet has the shortest wavelength in the visible spectrum.

3 On the first page of Student Sheet 6.2: Constructing the Spectrum are drawings of waves that represent red, green, and blue light. Identify each of these waves. Cut them out and attach each to the appropriate row in Table 1 on the second page of the student sheet.

4 Use a pencil to draw the waves for orange, yellow, indigo, and violet in the appropriate rows of Table 1.

5 Read "Infrared and Ultraviolet" to learn about waves that are shorter and waves that are longer than those of visible light.

6 Identify the wave in Table 1 that represents ultraviolet and the wave that represents infrared. Correctly label these waves on the table.

READING SELECTION

BUILDING YOUR UNDERSTANDING

COLOR AND WAVELENGTH

You have been modeling light waves using your chain. But what does this have to do with color? Our eyes can detect different wavelengths of light. Our eyes and brain perceive these different wavelengths as color. We cannot see all the wavelengths of electromagnetic radiation—some are invisible. We see only those wavelengths that range from about 400 to 800 nanometers (a nanometer is one billionth of a meter (1×10^{-9} m), so these wavelengths are very tiny). ∎

INFRARED AND ULTRAVIOLET

The spectrum of visible light is part of a much larger spectrum of waves called the electromagnetic spectrum. The electromagnetic spectrum contains wavelengths that are much longer and wavelengths that are much shorter than those of visible light. Most wavelengths of electromagnetic waves are invisible to the human eye. For example, infrared is invisible electromagnetic radiation, with wavelengths a little longer than red light. Ultraviolet is invisible electromagnetic radiation, with wavelengths slightly shorter than violet light. ■

REFLECTING
ON WHAT
YOU'VE DONE

1. Draw a wave in your science notebook. Label the following features on the wave: wavelength, amplitude, trough, and crest.

2. Answer the following questions in your science notebook, then discuss your answers with the class:

A. In the wave model for light, what feature of the wave determines how your eyes sense the color of a particular wave of light?

B. In the visible spectrum, your eyes detect the longest wavelength as what color?

C. In the visible spectrum, your eyes detect the shortest wavelength as what color?

D. Which type of electromagnetic radiation occurs directly beyond the red end of the visible spectrum?

E. Which type of electromagnetic radiation occurs directly beyond the violet end of the visible spectrum?

Burning Our Biggest Organ

Put on your hat, and don't forget the sunscreen. Yes, it's one of those hot summer days, ideal for having fun in or on the water. But spending the morning tubing on a river or swimming at the local pool could cause you to severely burn the largest organ in your body!

No, it's not your liver; it's your skin. The cells that make up the skin all work together to perform a specific function. The skin's job is to separate the inside of your body from the outside world, and the skin does its job well. Outside the body, there are many harmful things. One of these is electromagnetic radiation. Visible light is fairly harmless, but some types of invisible electromagnetic radiation can be very damaging. For example, ultraviolet radiation (UV) causes sunburn.

The Sun produces lots of UV in a variety of different wavelengths. Luckily, most UV is absorbed high in the atmosphere and never reaches Earth's surface. The layer of the atmosphere that absorbs UV is called the ozone layer (see "Ozone and the Ozone Layer" sidebar).

Tanning is the skin's natural reaction to UV. When skin tans, it produces a brown substance called melanin. Cells deep in the skin produce melanin, which acts as a protective barrier by absorbing harmful radiation. In light-skinned people, these cells are less active. People with darker skin produce more melanin than people with lighter skin. In fact, the darkness of people's skin is mainly an indicator of the amount of melanin they produce. Different skin colors are probably the result of adaptation to different levels of UV at different parts of the globe.

Sunburn is the body's reaction to damage caused by UV. Overexposure to UV can kill cells or damage parts of cells, including DNA. The body reacts as it would to an ordinary burn or other skin damage—but slower. A few hours after exposure to the Sun, the skin becomes red, inflamed, and painful. This is why it is easy to get sunburned and not know it until the following day!

A minor sunburn often heals after a few days. (A bad sunburn, like any burn, can kill.) But repeatedly burning one part of the skin can cause skin cancer. That's why it is important to protect your skin from UV. One way is simply to cover up—wear a wide-brimmed hat, long-sleeved shirt, and long pants.

▸ **DARK SKIN PROVIDES BETTER PROTECTION FROM UV LIGHT. DIFFERENT SKIN COLORS OFFER DIFFERENT LEVELS OF PROTECTION.**

PHOTO (upper right): National Science Resources Center
PHOTO (below): National Science Resources Center
PHOTO (lower right): Courtesy of Libby Crowe

READING SELECTION
EXTENDING YOUR KNOWLEDGE

OZONE AND THE OZONE LAYER

The ozone layer of Earth's atmosphere protects us from most of the Sun's UV radiation. Ozone—a type of oxygen—is able to absorb UV. A layer of ozone about 24 kilometers (15 miles) up in the atmosphere shields Earth. The amount of ozone present high in the atmosphere varies naturally over time. However, in the last 30 years scientists have found that human activity is slowly destroying the ozone layer. Pollution, such as some chemicals used in old aerosol cans, refrigerators, and air conditioners, gets into the ozone layer and reacts with and destroys ozone. Scientists have observed big "holes" appearing in the ozone layer where the levels of ozone are very low. The largest of these holes is in the Southern Hemisphere. Scientists also have observed declining levels of ozone in the ozone layer over the whole globe.

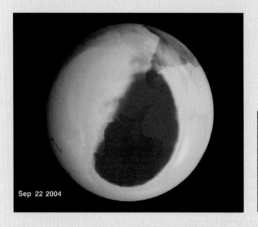

Sep 22 2004

▶ WE ARE PROTECTED FROM SOME UV BY THE OZONE LAYER. HOWEVER, POLLUTION IS REDUCING THE AMOUNT OF OZONE IN THE OZONE LAYER MAKING IT LESS EFFECTIVE. SCIENTISTS NOW KEEP WATCH ON A LARGE HOLE (SHOWN HERE AS A BLUE BLOB OVER ANTARCTICA) THAT WAS DISCOVERED IN THE OZONE LAYER OVER PART OF THE SOUTHERN HEMISPHERE.

PHOTO: NASA/Goddard Space Flight Center Scientific Visualization Studio

▶ THIS IS A MELANOMA. IT IS ONE FORM OF SKIN CANCER. MELANOMAS CAN BE CAUSED BY OVEREXPOSURE TO UV.

PHOTO: National Cancer Institute

Another way to get some protection from UV is to use a sunscreen. Sunscreens work in two ways. Some sunscreens look white; they simply reflect UV. Others absorb the UV. They work just like our natural skin protector, melanin. If you look at a bottle of sunscreen, you'll notice it is labeled with a Sun Protection Factor (SPF). The higher the SPF, the more effective the sunscreen.

So the next time you are out in the Sun, put on a hat and slather on the sunscreen! Your skin will thank you. ■

DISCUSSION QUESTIONS

1. Why has the problem of sunburn gotten worse over the past 50 years or so? What can we do about it?

2. Why are some people darker skinned than others? What advantage does it have with respect to sunlight?

▶ ARE THESE PEOPLE ENDANGERING THE BODY'S LARGEST ORGAN?

PHOTO: Phil Whitehouse/creativecommons.org

EXPLORATION ACTIVITY: INVESTIGATING THE INVISIBLE

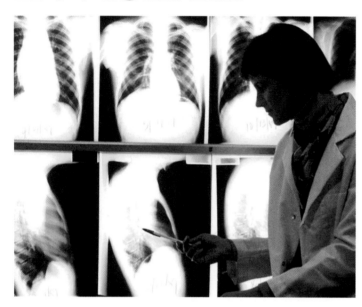

▶ **WHAT KIND OF ELECTROMAGNETIC WAVES MAKE PICTURES LIKE THIS POSSIBLE?**

PHOTO: National Cancer Institute

INTRODUCTION

In this lesson, you will begin the Exploration Activity. You will work on this activity over the next few weeks. What is the Exploration Activity? It is an extended project that gives you the opportunity to expand your knowledge and apply what you have learned in the unit to the world around you. In this Exploration Activity, you will select a region of the electromagnetic spectrum to examine in more detail. You will investigate the region's wavelengths and frequency characteristics, how the waves in that region are produced, and how we are able to use the waves in that part of the spectrum. You will use the library, the Internet, and other resources to conduct your research. You will then use the information you collect to make a presentation to the class (supported by a visual aid you have made) about the part of the spectrum that you select. The work you do for the Exploration Activity will be an important part of your grade for this unit. You will be given several homework assignments and a small amount of class time to do this work. However, you will do most of it on your own time. At the end of the unit, two to three class periods will be used for Exploration Activity presentations.

OBJECTIVES FOR THIS LESSON

Select a part of the electromagnetic spectrum to research.

Research the region of the electromagnetic spectrum you have chosen.

Create a visual aid to support an oral presentation about what you have learned.

Give an oral presentation on the region of the electromagnetic spectrum you researched.

▶ **MATERIALS FOR PART 1 BEGINNING THE EXPLORATION ACTIVITY**

For you

1   copy of Student Sheet 7.1: Exploration Activity Schedule

GETTING STARTED

1 What do you think is meant by the term "invisible"? Discuss your ideas with your group, then with the class.

2 How can something be invisible? Contribute to a class discussion about this idea.

3 Can you think of something that is invisible but that is useful in some way? Can you think of any others? As your class compiles a list of invisible things that we know about and that are useful, record the list in your science notebook. ✍

4 Read about infrared light in "The Hidden Spectrum" on pages 84-86. When you have finished, answer the following questions in complete sentences:

A. Who discovered infrared light?

B. Briefly describe the experiment done by the person who discovered infrared light.

C. Who discovered ultraviolet light?

D. What did these discoveries about infrared and ultraviolet light reveal about the electromagnetic spectrum?

5 Now read "Viewing the World and Beyond in Infrared" on pages 87-89. When you have finished reading this selection, list three ways that we use infrared light.

▶ **WHAT INVISIBLE FORCE IS ACTING ON THIS DOG?**

PHOTO: Dan Bennett/creativecommons.org

EXPLORATION ACTIVITY GUIDELINES

PART 1

BEGINNING THE EXPLORATION ACTIVITY: PLANNING YOUR RESEARCH

PROCEDURE

1 After your teacher gives you Student Sheet 7.1: Exploration Activity Schedule, tape it in the front of your science notebook. You will need to refer to it as you work on the Exploration Activity. Follow it carefully.

2 Record the due dates for each component on the schedule. Follow along as your teacher reviews the rest of the Exploration Activity Guidelines.

3 You (and your partner, if you are working in pairs) will choose a region of the electromagnetic spectrum to research. Look at the chart of the electromagnetic spectrum on page 85 and select a region of the spectrum that you would like to investigate.

4 Discuss your choice with your teacher. You may be asked to choose another region if another student pair has already chosen the same region. It is best if all regions of the electromagnetic spectrum are covered by the class.

5 The information you gather will be divided into three sections. As you gather information, write your notes under these headings:

DISCOVERY

- Who discovered the waves in this region of the electromagnetic spectrum?
- How were these waves discovered?

WAVE PROPERTIES

- What are the wavelengths of these waves? Try to relate the wavelength to something that you can see and measure.
- What is the frequency of these waves?
- How are these electromagnetic waves produced?
- What instruments are used to detect these waves?

WAVE USES

- How are these waves used? Find at least three uses. Include pictures showing how the waves are used.
- In what ways are these waves beneficial?
- Are these waves harmful in any way? If so, how? How can we protect ourselves from any harmful effects of these electromagnetic waves?

6 Review the questions in Step 5. If you are working with a partner, decide how you will divide the work. If you are working alone, develop a research plan.

PART 2

CONDUCTING YOUR RESEARCH

PROCEDURE

1. To help with planning your research, write an outline of your research (see Figure 7.1). Your outline should use the headings provided in Step 5 of Part 1.

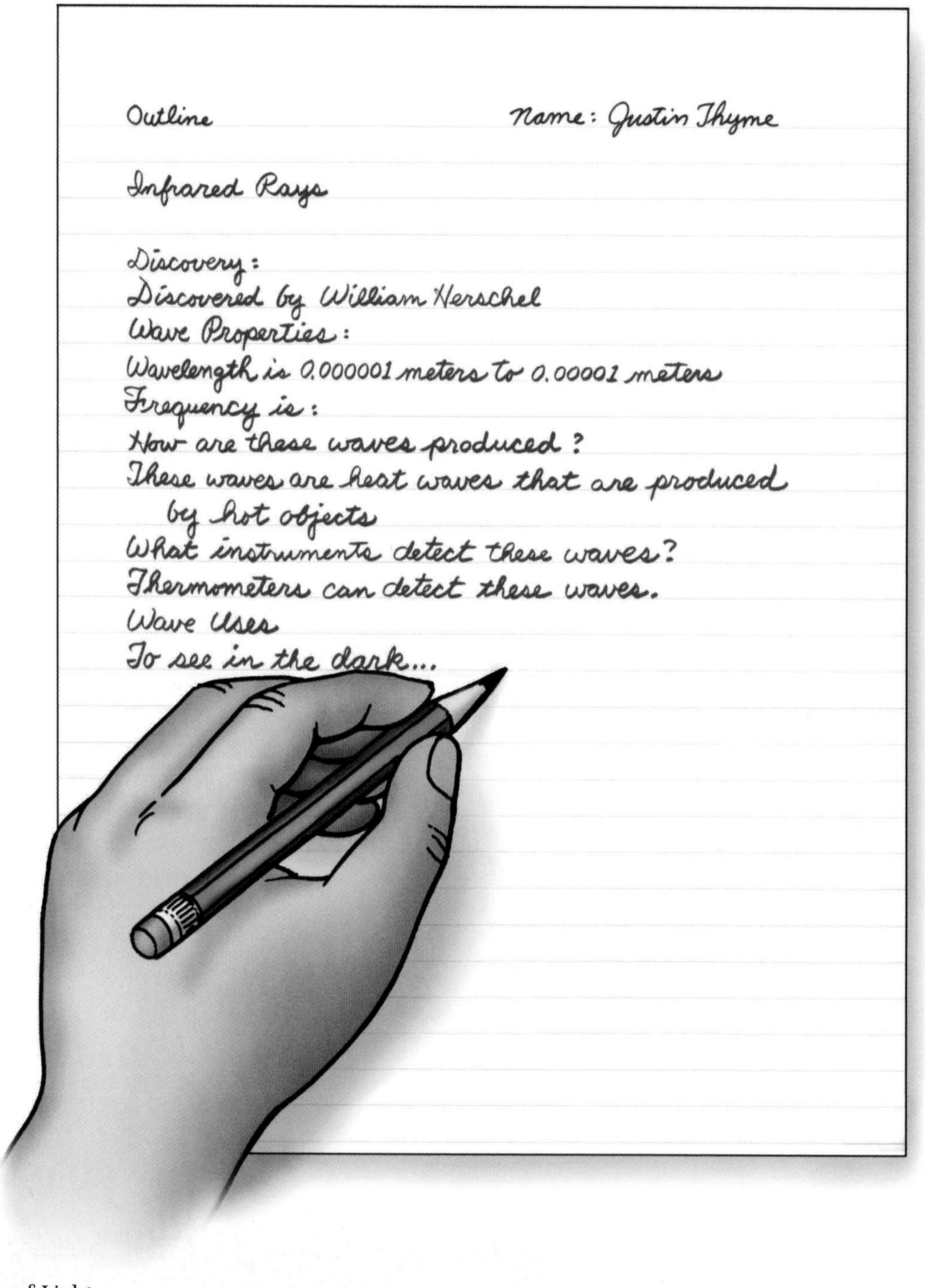

Outline   Name: Justin Thyme

Infrared Rays

Discovery:
Discovered by William Herschel
Wave Properties:
Wavelength is 0.000001 meters to 0.00001 meters
Frequency is:
How are these waves produced?
These waves are heat waves that are produced
   by hot objects
What instruments detect these waves?
Thermometers can detect these waves.
Wave Uses
To see in the dark...

A SAMPLE OUTLINE FOR THE EXPLORATION ACTIVITY. THE ELECTROMAGNETIC WAVE CHOSEN WAS INFRARED.
FIGURE **7.1**

AN EXAMPLE OF A
BIBLIOGRAPHY WRITTEN
FOR THE EXPLORATION
ACTIVITY. YOUR TEACHER
MAY SHOW YOU A
DIFFERENT WAY TO WRITE
YOUR BIBLIOGRAPHY.
FIGURE **7.2**

Name: Justin Thyme

Bibliography

Toler, D. and David Marsland. Reflections on Light. Washington, D.C. NSRC Publications. 2008

Gribbon, John. A Brief History of Science. Barnes and Noble Books. 2005

Burnie, David. Light. Dorling, Kindserley Publishing, Inc. 1998

2 As you collect information, record your sources in a bibliography. The bibliography can include sources such as books, CD-ROMs, DVDs, videotapes, TV programs, and magazines. It should include at least one website and one book (other than an encyclopedia). A sample bibliography is shown in Figure 7.2.

3 Hand in your outline and a copy of your bibliography on separate sheets of paper by the due date on the schedule. Your teacher will use this information to help you make sure your research is heading in the right direction.

PART 3

PRESENTING WHAT YOU'VE LEARNED

▶ **MATERIALS FOR PART 3 PRESENTING WHAT YOU'VE LEARNED**

For you

1 copy of Student Sheet 7.2: Points to Consider When Designing and Preparing a Visual Aid

1 copy of Student Sheet 7.3: Scoring Rubric for the Exploration Activity Presentation

PROCEDURE

1 Prepare your presentation. You will give a 3-minute oral presentation supported by at least one visual aid. Your visual aids can be a poster, web page, computer presentation, or model. Your teacher will outline some points to think about when making your visual aid.

2 Design your visual aid carefully. Refer to Student Sheet 7.2 as you work to avoid pitfalls.

3 Create your visual aid.

4 Prepare and practice your oral presentation. Ask your parents, teacher, or friends to suggest ways to improve your presentation. Follow these suggestions:

A. If working with a partner, make sure you design your presentation so that you both contribute equally to the presentation.

B. Refer frequently to the visual aid as you give your presentation.

C. If you get nervous when speaking or have a bad memory, make short notes (not complete sentences) on index cards to use during your presentation.

D. Make sure your voice can be heard clearly in the back of the room.

5 Student Sheet 7.3 shows the rubric your teacher will use to score your presentation. Use this rubric to help you plan your presentation so that you earn a high score.

6 Your teacher will tell you the day on which you will make your presentation. Be sure you are prepared.

7 You will have only 3 minutes to make your presentation.

8 Speak clearly. If you are working with a partner, make sure you both contribute equally to the presentation. Don't forget to refer to your visual aid as you make your presentation.

9 After your presentation, hand in your visual aid and the final version of your bibliography.

REFLECTING
ON WHAT
YOU'VE DONE

1 After all the presentations have been made, discuss with your group which was the most interesting and which was presented best. Try to identify the reasons for your choice(s).

2 How do you think you could have improved your presentation? Make a short list of these improvements in your science notebook.

▶ THINK CAREFULLY ABOUT HOW YOU DESIGN YOUR VISUAL AID. WHEN MAKING A PRESENTATION, USE YOUR VISUAL AID TO HELP YOU EXPLAIN WHAT YOU LEARNED ABOUT YOUR REGION OF THE ELECTROMAGNETIC SPECTRUM.
FIGURE **7.3**

THE HIDDEN SPECTRUM

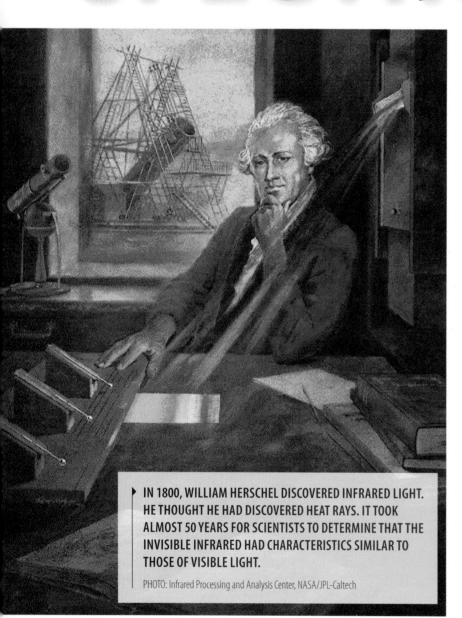

▶ IN 1800, WILLIAM HERSCHEL DISCOVERED INFRARED LIGHT. HE THOUGHT HE HAD DISCOVERED HEAT RAYS. IT TOOK ALMOST 50 YEARS FOR SCIENTISTS TO DETERMINE THAT THE INVISIBLE INFRARED HAD CHARACTERISTICS SIMILAR TO THOSE OF VISIBLE LIGHT.

PHOTO: Infrared Processing and Analysis Center, NASA/JPL-Caltech

In 1800, the famous astronomer William Herschel (1738-1822) began to study sunspots. To protect his eyes, he added special filters to his telescope. He pointed the telescope at the Sun and was able to see sunspots very clearly.

But after a few minutes, he ran into a problem. His telescope got too hot to use. Why was this happening? Was there a link between the temperature of his telescope and light? He decided to conduct some experiments to try to answer these questions.

In one experiment, Herschel tried to find out which color in white light produced the most heat when it hit a surface. He used a prism to split white light into its different colors. He put a thermometer in each line of colored light and measured the increase in temperature produced by each color. He discovered that red light produced the most heat. He then put a thermometer just outside the red end of the spectrum. Here no light was visible, but to his surprise the thermometer got even hotter. He had discovered an invisible form of light. Herschel incorrectly thought he had discovered heat rays. This invisible light was later called infrared. Were there other types of invisible light? The search was on!

WIDENING THE SEARCH

Johann Wilhelm Ritter (1776–1810), a German scientist, heard about Herschel's discovery. He decided to use a light-sensitive chemical to search for other types of invisible light. He used the chemical to test for light at the violet end of the visible spectrum. He found another invisible form of light and named it ultraviolet.

Soon the idea of a whole range, or spectrum, of radiation began to grow. Scientists realized that visible light made up only a tiny part of a much wider spectrum. The idea of a spectrum fit neatly with the emerging idea that light was a form of wave energy. Different colors and types of light belonged to the same spectrum. They simply had different wavelengths. Later scientists discovered that the waves that make up the spectrum are electromagnetic waves.

Many types of electromagnetic waves have been identified. Wavelengths can be as long as tens of kilometers (radio waves) and as short as 0.000000000000001 millimeters—10^{-15} meters—(gamma rays). ■

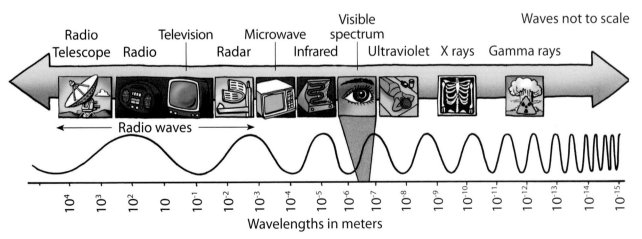

▶ THE DIFFERENT WAVELENGTHS OF THE ELECTROMAGNETIC SPECTRUM HAVE A VARIETY OF NAMES AND USES.

READING SELECTION

EXTENDING YOUR KNOWLEDGE

TUNING IN

Have you ever tried tuning a radio? Have you wondered about those numbers on a digital display or scale? And what's all this about AM and FM?

The numbers refer to the frequency of the electromagnetic waves—radio waves—you are tuning into. AM and FM refer to groups, or bands, of radio frequencies. AM frequencies are around 1 million waves per second or 1 megahertz (MHz). Waves in the FM band are about 100 times shorter, with a frequency of 100 MHz. Remember, the shorter the wave, the higher the frequency (the number of waves that pass a certain point every second).

▶ WHICH FREQUENCY DO YOU TUNE INTO MOST FREQUENTLY?

DISCUSSION QUESTIONS

1. How did Herschel's discovery expand our understanding of the electromagnetic spectrum? How did it inspire other scientists?

2. How has an understanding of electromagnetic spectrums affected our daily lives?

VIEWING THE WORLD AND BEYOND IN INFRARED

Infrared may be invisible, but it can still be detected. Hot objects give off infrared. Hold your hands near a heater and they will soon warm up. When the infrared released by the heater strikes our hands, it is transformed back into heat. By using special film or special electronic cameras, you can see the world in infrared. One such device could be looking at your school right now. Satellites in orbit around the Earth are designed to view the planet in infrared. The images they collect can provide information on the condition of crops, pollution, or the temperature of the oceans.

Weather satellites take images in infrared. Meteorologists use these images to record the temperature of air masses moving across Earth.

These images also can record the temperature of the clouds inside giant storms.

Infrared detectors in space also look away from Earth. Many regions of the universe cannot be seen with optical telescopes. These regions are hidden by clouds of gas and dust. However, infrared can pass through these clouds. So it is possible to see through these clouds by using orbiting infrared observatories such as the Infrared Astronomical Satellite (known as IRAS). IRAS gives astronomers a different perspective on the universe. With it, they can look into the center of our galaxy. Infrared can be used to detect objects much cooler than stars. Astronomers also use infrared to look for planetary systems other than our own.

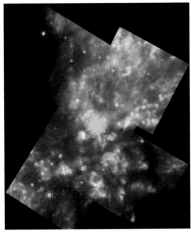

▶ **VISIBLE LIGHT PHOTOS OF THE ORION NEBULA LOOK LIKE A GLOWING CLOUD CONTAINING A FEW STARS, AS ON THE LEFT. THE PICTURE ON THE RIGHT WAS TAKEN USING INFRARED WAVELENGTHS FROM PART OF THE ORION NEBULA AND REVEALS MUCH MORE.**

PHOTOS: NASA, R. Thompson and S. Stolovy,(University of Arizona), and C.R. O'Dell (Rice University).

READING SELECTION
EXTENDING YOUR KNOWLEDGE

▶ THIS IMAGE SHOWS THE RAINFOREST IN RONDONIA, BRAZIL. IT WAS CREATED USING NEAR-INFRARED, RED, AND GREEN LIGHT. THE IMAGE CAN BE USED TO IDENTIFY AREAS WHERE THE RAINFOREST HAS BEEN CUT DOWN, WHICH APPEAR PALE RED OR BROWN.

PHOTO: Courtesy of NASA/GSFC/METI/ERSDAC/JAROS and U.S./Japan ASTER Science Team

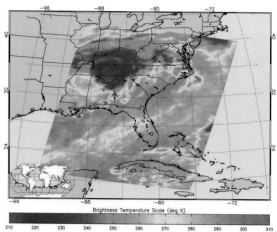

▶ RAINFOREST DESTRUCTION, SUCH AS THAT CAUSED BY FIRES, CAN BE MONITORED USING EQUIPMENT ON EARTH-ORBITING SATELLITES. SOME OF THESE SATELLITES LOOK AT EARTH USING INFRARED DETECTORS. THE RED DOTS IN THIS IMAGE OF THE RAINFOREST IN CENTRAL SOUTH AMERICA REPRESENT ACTIVE FIRES.

PHOTO: NASA image by Jeff Schmaltz, MODIS Rapid Response Team, Goddard Space Flight Center

▶ IN THIS INFRARED IMAGE OF HURRICANE KATRINA, THE ORANGE AND RED AREAS OF THE STORM REPRESENT WARMER TEMPERATURES IN CLOUD-FREE AREAS.

PHOTO: NASA Jet Propulsion Laboratory

USING INFRARED CLOSER TO HOME

Back here on Earth, seeing in infrared gives us a new view of our world—even a new view of ourselves. Look at the human face in infrared. You can see the different temperatures of skin. The human face begins to look quite different, doesn't it?

Looking at objects in infrared provides engineers with useful information. For example, they can use infrared detectors to discover how buildings lose heat. This helps them design buildings that are more energy efficient. They also can use the detectors to check electrical systems for overheating that could cause fires. Computer makers use infrared sensors to look for hot spots on circuit boards. Automakers use these sensors to make sure that car engines are running properly. Railroads use infrared sensors to spot overheated wheels on railroad cars.

By making the invisible visible, infrared photography and detection allow people to see a world that previously was hidden from view. ■

DISCUSSION QUESTIONS

1. How is infrared radiation used by scientists and by others?

2. Besides infrared detection, how else has technology helped us to see "a world that was hidden from view"?

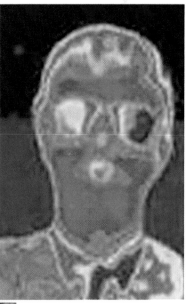

▶ THIS INFRARED IMAGE USES COLOR TO SHOW WHERE INFRARED IS GIVEN OFF. THE RED AREAS SHOW PARTS OF THE FACE GIVING OFF THE MOST INFRARED. THIS IMAGE SHOWS A HOT FACE BUT COOL SUNGLASSES!

PHOTO: Infrared Processing and Analysis Center, NASA/JPL-Caltech

▶ AN INFRARED PHOTOGRAPH SHOWS THAT THIS HOUSE HAS A WELL-INSULATED ROOF (BLUE-GREEN AREAS), BUT LOSES MUCH OF ITS HEAT THROUGH ITS WALLS AND WINDOWS (ORANGE AREAS).

PHOTO: Courtesy of DOE/NREL, Credit—Paul Torcellini

LOOKING AT COLORS

INTRODUCTION

In Lesson 5, you used your prism to observe a spectrum when white light passed through the prism. You saw that white light is a combination of colored lights. Why do many objects appear to be only one color when illuminated by white light? Why does the color of these objects appear to change when you look at them through transparent colored glass or plastic? In this lesson, you will use prisms and your eyes to try to find the answers to these questions.

▶ **WHY DO THE FLOWERS IN THE PICTURE APPEAR TO BE COLORED? WHAT COLOR WOULD THEY BE IF THERE WERE NO LIGHT?**

PHOTO: Bruce Fritz, Agricultural Research Service/U.S. Department of Agriculture

OBJECTIVES FOR THIS LESSON

Examine some colored objects and explain why they are colored.

Use colored sheets to examine a spectrum produced from white light.

Predict and observe the appearance of different colors when they are viewed through transparent colored sheets.

▶ **MATERIALS FOR LESSON 8**

For you

   1   copy of Student Sheet 8.1: Looking at the Spectrum Through Transparent Colored Sheets

   1   copy of Student Sheet 8.2: Looking at Colors Through Filters

For your group

   1   triangular prism
        Transparent colored sheets (red, green, and blue)
        Colored pencils (red, green, and blue)

For your group and another group to share

   1   ray box
   1   ray box lid
   1   60-W clear halogen lightbulb
   1   extension cord
   1   bulb holder
   2   narrow-slit ray box masks
   2   no-slit ray box masks

GETTING STARTED

1 One member of your group should collect the materials as directed by your teacher.

2 Look at the transparent colored sheets and colored pencils you have been given. Discuss with your group why these objects are colored. Be prepared to share your ideas with the class.

▶ **TO OUR EYES, THE WORLD IS A COLORFUL PLACE.**

PHOTO: Kathleen Tyler Conklin/www.flickr.com/photos/ktylerconk

LOOKING AT THE SPECTRUM THROUGH TRANSPARENT COLORED SHEETS

PROCEDURE

1 Use the ray box and prism as you did in Lesson 5 to create a spectrum from white light. Now look at the spectrum through each of the colored sheets. Record your observations on Student Sheet 8.1: Looking at the Spectrum Through Transparent Colored Sheets in a table of your own design.

2 Discuss your observations with your group, then write a short paragraph explaining your observations. Below are some questions you might consider:

A. Is the spectrum observed through the colored sheets the same as the spectrum for white light?

B. Are more or fewer colors visible when you look through the colored sheets?

C. What do the colored sheets do to white light?

3 Sometimes transparent colored sheets are called color filters. Why do you think that is? Write an explanation.

4 Write a short answer to the following question: What do red filters do to white light?

INQUIRY 8.2

LOOKING AT COLORS THROUGH FILTERS

PROCEDURE

1. Select the red, green, and blue colored pencils. In the first column of Table 1 on Student Sheet 8.2: Looking at Colors Through Filters, make a swatch of each color. White and black have been done for you. Label each swatch with the name of the color—use the color name written on the side of each colored pencil.

2. Look at the color swatches through the red filter only. Hold the filter about 10 cm above the swatch, as shown in Figure 8.1. Do not place it on the paper. Record the appearance of each color in the "Red" column of Table 1.

▶ LOOK THROUGH THE RED FILTER AT THE COLOR SWATCHES IN TABLE 1. HOLD THE RED FILTER ABOUT 10 CM ABOVE THE SURFACE OF THE PAPER.
FIGURE **8.1**

3 Predict how the same colors will appear under the green filter. Record your predictions in Table 1.

4 Test your predictions and record the results.

5 Repeat the same procedure using the blue filter.

6 Discuss your observations with your group and then with the class. In writing, describe any patterns you notice in your results. Can you explain these patterns? Record any ideas you may have.

REFLECTING
ON WHAT
YOU'VE DONE

❶ Review your explanations and observations on Student Sheet 8.2 before drawing the following diagrams in your science notebook.

A. Draw a diagram that explains why a green object appears green in white light.

B. Draw a diagram that explains the appearance of an object that is green when viewed through a red filter.

❷ What will you observe if white light passes through both a red filter and a blue filter?

A. Record your prediction and observations in your notebook, then write a paragraph to explain your observations.

❸ Read "Why Objects Look Colored" on pages 96-98. Think about the ideas you had at the beginning of the lesson about why objects were colored. Have these ideas changed?

why OBJECTS look COLORED

You can see objects because light travels from an object to your eyes. Objects you can see are said to be visible. Light sources—luminous objects, such as the Sun or a fluorescent lightbulb—are visible because they make visible light. But opaque, nonluminous objects do not make their own light. They are visible to your eyes because light from light sources bounces off them. In other words, light reflects off them.

Why do objects that reflect light appear to be colored? What, for example, happens when white light (made up of all colors of the visible spectrum) hits a yellow flower? The flower absorbs some colors in the white light and some—those that make up yellow—are reflected. You see this yellow light when some of it is reflected to your eyes.

Take another example. A red car looks red in white light because the car reflects only red light. The paint on the car absorbs the other colors. Most objects reflect some colors or wavelengths of light and absorb others. Only silvered mirrors or white objects reflect all colors. Only completely black objects absorb all the colors of the spectrum. No light reflects from the surfaces of completely black objects, so no light from black objects reaches your eyes.

▶ THE LENSES OF SUNGLASSES ARE OFTEN GRAY. THEY REDUCE THE INTENSITY OF SUNLIGHT BY FILTERING OUT SOME OF THE LIGHT. GRAY LENSES ARE USED SO THAT ALL COLORS THE EYES DETECT ARE REDUCED EQUALLY. THE SUNGLASSES PRODUCE A SCENE THAT IS DIMMER BUT STILL NORMALLY COLORED.

PHOTO: Eric Long, Smithsonian Institution

Some colors may also be produced by the reflection of more than one wavelength of light. However, your eyes detect these wavelength mixtures as only one color. This concept may seem confusing, but it is something you have already discovered. You see the paper of this page as white, but you know that the light reflected from the paper to your eyes contains all the colors of the visible spectrum.

FILTERING LIGHT

There is a slightly different explanation for why an object observed through a filter appears to be the color you see. Filters allow only some wavelengths of light through them—they transmit some colors and remove (absorb) others—in other words, filters filter out some colors (or wavelengths). So when you look at a sheet of white paper through a red filter, the paper appears red. It looks red because the filter absorbs all the colors except red from the white light the paper reflects. Only red light reflecting off the paper reaches your eyes. If you look at a blue object through the same filter, the object appears black. This is because the blue object reflects only blue light; none of the blue light reflected by the blue object can pass through the red filter. Therefore no light from the object reaches your eyes—the object appears to be black.

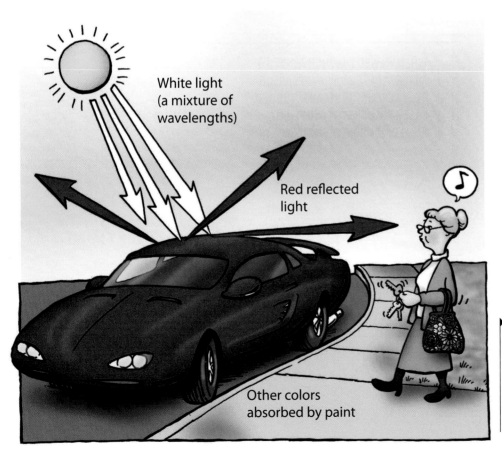

White light (a mixture of wavelengths)

Red reflected light

Other colors absorbed by paint

▶ WHY DOES THIS CAR LOOK RED? THE PAINT ON THE CAR ABSORBS MOST OF THE WAVELENGTHS IN WHITE LIGHT AND REFLECTS ONLY THOSE WAVELENGTHS THAT LOOK RED WHEN DETECTED BY OUR EYES.

READING SELECTION

EXTENDING YOUR KNOWLEDGE

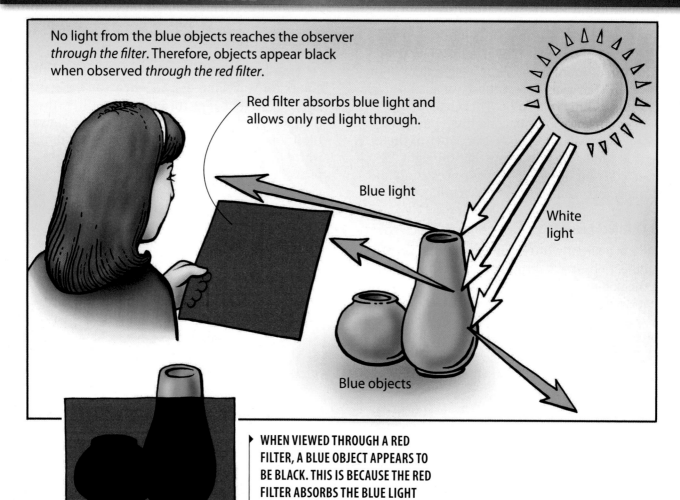

No light from the blue objects reaches the observer *through the filter*. Therefore, objects appear black when observed *through the red filter*.

Red filter absorbs blue light and allows only red light through.

Blue light

White light

Blue objects

What the girl sees

▶ WHEN VIEWED THROUGH A RED FILTER, A BLUE OBJECT APPEARS TO BE BLACK. THIS IS BECAUSE THE RED FILTER ABSORBS THE BLUE LIGHT REFLECTED FROM THE OBJECT.

Objects or filters look colored because they absorb or subtract some colors from white light. This process of producing color by subtracting colors or wavelengths from white light is called color subtraction. When you paint a picture in art class or look at a colored illustration printed in this book, the color you see has been produced by color subtraction. In the next reading selection, you will learn more about how people developed the techniques of color printing. ∎

DISCUSSION QUESTIONS

1. How does the color that objects appear depend on whether they absorb light or reflect light?

2. Why do color filters change the color of an object viewed through the filter?

PRINTING in COLOR

Try to imagine a world without printed books, newspapers, or magazines. For many thousands of years, people have used pictures and writing to communicate knowledge and ideas. But for most of this time, writing and pictures were not printed; they were copied by hand.

Buddhist monks in ancient China were the world's earliest expert printers. Over one thousand years ago, they used hand-carved blocks of wood to make thousands of copies of holy texts. The blocks were inked and then pressed onto silk and paper. The ancient Chinese were also the first to experiment with color in printed books.

While the Chinese already had developed their printing techniques, the Europeans were still hand-copying most of their books. Monks did most of this painstaking work. They then added color by hand in a process called illumination.

▶ IN MEDIEVAL EUROPE, BOOKS USUALLY HAD TO BE COPIED AND COLORED BY HAND.

PHOTO: Library of Congress, Prints & Photographs Division, LC-USZ62-110307

READING SELECTION
EXTENDING YOUR KNOWLEDGE

MECHANIZING THE PRINTING PROCESS

More than a thousand years after the invention of printing, the process became mechanized. The printing press was invented.

We don't know who invented the first printing press, but we do know that printing quickly became big business. In the 1450s, a German printer, Johannes Gutenberg, invented a method of making metal letters (called type) and casting individual metal letters in standard-sized molds. This sped up the printing process. It became much cheaper to print books and pamphlets. However, pictures were still hand-carved on wooden blocks or metal plates. These pictures were black and white, but were sometimes colored by hand.

As books became more available, more people learned to read. The demand for books increased. People wanted to read about far-off places and exciting new discoveries. They wanted their books to have more pictures, and they wanted these pictures to be colored. This drove the development of color printing.

▶ THIS PICTURE IS FROM A BOOK BY THE NATURALIST JOHN JAMES AUDUBON (1785–1851). EARLY EDITIONS OF AUDUBON'S BOOKS CONTAINED HAND-TINTED PICTURES. ILLUSTRATIONS FROM LATER EDITIONS USED AN IMPROVED TYPE OF LITHOGRAPHIC COLOR PRINTING CALLED CHROMOLITHOGRAPHY.

PHOTO: Library of Congress, Prints & Photographs Division, LC-USZC4-722

COLOR FOR EVERYONE

The first successful method of large-scale color printing (called lithography) was invented near the end of the 19th century. Originally in lithography, designs were drawn on fine-grained stone with a special oily crayon or ink. The stone was then moistened with water. Because water and oil repel each other, the oily picture did not get wet—only the rest of the stone. Oily ink was then rolled onto the wet stone. Again because water and oil repel each other, the ink stuck only to the oily picture, not to the surrounding wet part of the stone. Paper was then pressed against the stone to make a print of the picture.

To print in different colors, multiple stones were used. One stone was used for each color. The paper had to go through the press many times to be pressed against each inked stone. The stones had to be carefully aligned so that the colors were printed in the correct place.

This process was faster than the illumination process done by hand, but it was still slow. At first, books printed by lithography were very expensive. However, with improved techniques and mechanization, the process eventually became inexpensive enough to print magazines and advertising circulars in color.

THE BEGINNING OF MODERN PRINTING

▶ THE PAGE SPREAD SHOWN IS FROM A VERSION OF A BIBLE PRODUCED BY GUTENBERG.

▶ JOHANNES GUTENBERG LED A REVOLUTION IN PRINTING USING MOVABLE TYPE THAT MADE THE PROCESS CHEAPER AND FASTER. PRINTING BECAME ONE OF THE FIRST MASS-PRODUCTION PROCESSES. HERE, GUTENBERG EXAMINES A PAGE FROM HIS IMPROVED PRINTING PRESS.

READING SELECTION

EXTENDING YOUR KNOWLEDGE

▶ IN MODERN PRINTING, FOUR PLATES, EACH REPRESENTING A DIFFERENT COLOR—CYAN, MAGENTA, YELLOW, AND BLACK— ARE MADE. USE A MAGNIFIER TO LOOK AT EACH PICTURE. YOU WILL SEE THEY ARE MADE FROM DOTS OF THE FOUR DIFFERENT COLORS. WHEN THESE COLORS ARE COMBINED, THEY PRODUCE A FULL-COLOR PICTURE.

PHOTO: DoD photo by TSGT Bob Fehringer

PRINTING WITH DOTS

Modern color printing is based on a similar process. These days, a picture is scanned electronically and broken into four separate images. Each image represents one of the colors— cyan, magenta, yellow, and black—that will make up the picture. The scanner also breaks up each colored image into dots. The more numerous the dots of a particular color, the stronger that color. Each image is then photographically transferred to a metal (not stone) printing plate.

On the printing press, the paper passes under each of the four plates. Each plate holds a different color of ink. Without a magnifier, our human eyes cannot see the tiny dots that make up a picture. The colored dots all seem to blend together—just as they would appear if you mixed them from colored paints. But if you look at the picture with a powerful magnifier, you will see these colors break down into the four colors used to print them. ■

DISCUSSION QUESTIONS

1. What are some key innovations that led up to the method of printing we use today?

2. Research the art movement pointillism and explain how it uses the same principles as a modern printer.

a GReeN eNGiNe
DRiVeN BY THe
sun

a giant snake slides silently across the wet floor of the Amazon rainforest. It's a 6-meter-long (19.7-foot-long) anaconda—the largest snake in world—and it is searching for a meal. Perhaps it will find a juicy capybara, a sort of giant guinea pig. If it's lucky, it may catch an unwary sloth on one of the sloth's rare visits to the forest floor.

The anaconda's method of capture is to squeeze its victim in the giant coils of its body until the victim stops breathing. But all this slithering and crushing requires the snake to use its giant muscles. These muscles—and the snake's other life processes—use energy. The snake gets its energy from food—perhaps today it will be the sloth. Of course, the snake never stops to think where the energy in food comes from. Do you?

Above the hunting anaconda is a canopy of green formed by the giant forest trees. Viewed from above, the forest stretches out in all directions—green as far as the eye can see. Could all this green have something to do with how the snake gets its energy? In fact, this rainforest is an enormous food factory powered by sunlight.

READING SELECTION

EXTENDING YOUR KNOWLEDGE

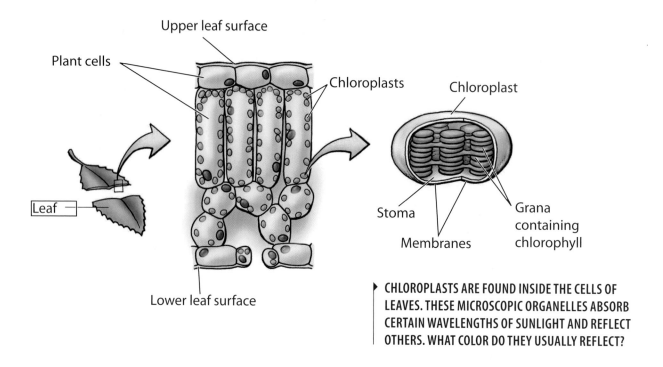

▶ CHLOROPLASTS ARE FOUND INSIDE THE CELLS OF LEAVES. THESE MICROSCOPIC ORGANELLES ABSORB CERTAIN WAVELENGTHS OF SUNLIGHT AND REFLECT OTHERS. WHAT COLOR DO THEY USUALLY REFLECT?

A green chemical in leaves called chlorophyll harnesses the light energy from the Sun. Chlorophyll absorbs some wavelengths of sunlight and reflects others. The forest canopy looks green because its leaves reflect green light. Chlorophyll uses some of the light energy it absorbs to drive chemical reactions. It transforms light energy into chemical energy—food. This food-making process is called photosynthesis, which means "making with light."

Inside each leaf cell are special tiny structures called chloroplasts, which contain chlorophyll. The flat leaves and the arrangement of chloroplasts in the cells are adapted to absorb sunlight. But plants need more than just chlorophyll and sunlight to make food by photosynthesis. Giant roots collect water from the shallow forest soil. Leaves absorb carbon dioxide from the air.

Each tree battles its neighbor for the resources needed for photosynthesis. The trees struggle upward, competing against each other for light. Their huge roots and trunks support the branches and leaves of the canopy, and also act as a highway for water and nutrients needed by the chloroplasts. Inside each chloroplast, the energy from the Sun is used to combine water and carbon dioxide (CO_2) to make sugars and starch. Each tree uses these substances to provide energy for its life processes. It also uses them to make more leaves, branches, flowers, and fruits. These leaves and fruits will in turn provide food energy for the sloth and the capybara.

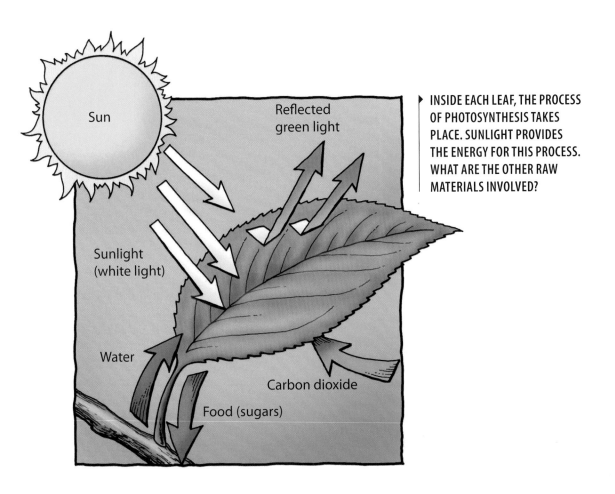

Sun

Reflected
green light

Sunlight
(white light)

Water

Carbon dioxide

Food (sugars)

INSIDE EACH LEAF, THE PROCESS OF PHOTOSYNTHESIS TAKES PLACE. SUNLIGHT PROVIDES THE ENERGY FOR THIS PROCESS. WHAT ARE THE OTHER RAW MATERIALS INVOLVED?

▶ THE ANACONDA IS THE WORLD'S LARGEST SPECIES OF SNAKE. IT USES POWERFUL MUSCLES TO SQUEEZE ITS PREY TO DEATH. THESE MUSCLES USE ENERGY WHEN THEY CONTRACT. WHERE DOES THIS ENERGY COME FROM?

PHOTO: Phil Whitehouse/creativecommons.org

▶ THIS SLOTH SPENDS MOST OF ITS LIFE IN THE TREES OF THE AMAZON RAINFOREST EATING LEAVES, COMING DOWN TO THE GROUND ONLY OCCASIONALLY. THE ENERGY IN ITS FOOD COMES FROM THE SUN. SLOTHS ARE OFTEN SLIGHTLY GREEN. THAT'S BECAUSE MICROSCOPIC PLANTS CONTAINING CHLOROPHYLL LIVE IN THEIR FUR.

PHOTO: davida3/creativecommons.org

The unsuspecting sloth chews leaves as it carefully descends a tree trunk to the forest floor. The capybara feeds on fruit recently fallen from the trees. The trees of the rainforest canopy provide the sloth with food energy. But this canopy of trees also blocks out the sunlight. In the semi-darkness of the forest floor there is little protection from the anaconda. Nearby, the snake, hunting by its sense of smell, tastes the air with a forked tongue.

Suddenly, the snake seizes the sloth and wraps its long body around it. The snake has found the energy—food—it needs to sustain itself. It doesn't realize that this food energy has come indirectly from the Sun, high above the forest canopy, through leaves of the forest to the leaf-eating sloth. The snake never sees the source of its energy. The Sun is hidden by the food factory of the forest canopy. ■

▶ VIEWED FROM ABOVE, THE GREEN RAINFOREST CANOPY HIDES THE FOREST FLOOR. WHY IS THE CANOPY GREEN? WHAT IS ITS FUNCTION? THESE ANSWERS LIE IN THE WAY CHLOROPHYLL ABSORBS AND REFLECTS SUNLIGHT.

PHOTO: Steph McGlenchy/creativecommons.org

DISCUSSION QUESTIONS

1. Make a diagram showing how various foods you eat are directly or indirectly dependent on solar energy.

2. Many scientists believe the extinction of dinosaurs was caused by the collision of a huge asteroid with Earth, making dust clouds that blocked sunlight for months. What chain of effects would you expect from the lack of sunlight for many months or years?

COLORED LIGHT

INTRODUCTION

Imagine you're at a play or a musical. The lights of the theater darken for the performance. A hush descends over the audience as the curtains slide open. The stage looks dark. Suddenly there is a blaze of color as the star of the performance makes her entrance bathed in colored light. Where is this light coming from? Above, and to either side of the stage, you locate the source of this light. Large numbers of spotlights—some fixed to gantries, a few directed by hand—point at the stage. A few emit white light, but most shine a beam of colored light onto the stage. How is this colored light produced? How are the different

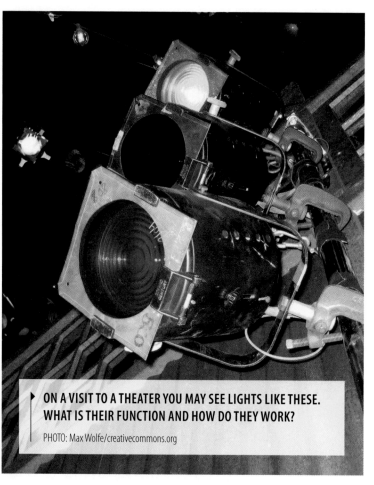

▶ ON A VISIT TO A THEATER YOU MAY SEE LIGHTS LIKE THESE. WHAT IS THEIR FUNCTION AND HOW DO THEY WORK?

PHOTO: Max Wolfe/creativecommons.org

colors used to set the scene on the stage combined? What happens when colored lights are mixed? How do these colors affect the appearance of the props and costumes? In Lesson 8, you determined that colored objects look colored because they absorb—or subtract—some of the colors from the light that falls on them and reflect other colors. Objects appear to be the color of the colors they reflect. What name did we give to this type of color mixing? In this lesson, you will investigate another type of color mixing. This color mixing process takes place when the colors from colored lights are added together. Can you guess the name that is given to this type of color mixing?

OBJECTIVES FOR THIS LESSON

▶ Investigate what happens when lights of different colors are mixed together.

▶ Discuss how this type of color mixing is used.

▶ **MATERIALS FOR LESSON 9**

For you

1 copy of Student Sheet 9.1: Mixing Colored Lights

For your group

3 flashlights
6 D-cell batteries
1 red filter
1 green filter
1 blue filter

GETTING STARTED

1. Imagine you are in charge of lighting the stage for a concert or play. You have red, green, and blue spotlights. With your group, predict which colors you could make by combining the colors of the spotlights.

2. Record your predictions in the Predictions section of Table 1 on Student Sheet 9.1: Mixing Colored Lights. Be prepared to share your predictions with the class.

▶ **HOW MANY COLORS OF LIGHT WILL YOUR GROUP PRODUCE?**

PHOTO: National Science Resources Center

MIXING COLORED LIGHTS

PROCEDURE

1 One member of your group should obtain the materials.

2 Working with your group, come up with a procedure to use the materials available to test the predictions you made in "Getting Started." Record the materials you use and your procedure in Table 1 on Student Sheet 9.1: Mixing Colored Lights. Use diagrams where appropriate.

3 Design a data table in Table 1 for your results.

4 Test your predictions and record your results.

5 In the Conclusions section of Table 1, write a paragraph explaining what you can conclude from your inquiry.

REFLECTING
ON WHAT
YOU'VE DONE

1 The type of mixing you explored in Inquiry 9.1 is called additive color mixing. In additive color mixing, red, green, and blue are the primary colors. Discuss the questions below with your group and then record your answers in your science notebook.

A. Why is the term "primary colors" used to describe red, green, and blue colors in additive color mixing?

B. Why do you think this type of mixing is called additive color mixing?

2 In Lesson 8, you learned about color mixing that subtracts color—subtractive color mixing. Write a short paragraph that describes how additive color mixing differs from subtractive color mixing.

3 Read "About Color Vision and Color Mixing" on page 112.

4 Review the question bank from Lesson 1. Are there more questions you can answer?

READING SELECTION

BUILDING YOUR UNDERSTANDING

ABOUT COLOR VISION AND COLOR MIXING

When you used a prism or spectroscope to split white light into its different wavelengths, the colors you observed were different wavelengths of light. Yellow, for example, was a small range of wavelengths that looked yellow.

Because of the way our eyes detect color, we perceive mixtures of colors as one color. For example, you already know that white is a mixture of colors.

The yellow you made with your flashlights is another example of this. It was not one wavelength of light. It was made from a mixture of two other colors—red and green—that your eyes detected and your brain perceived as being yellow. So the brain perceives yellow in two ways: pure yellow wavelengths or mixtures of red and green wavelengths. ■

▶ A. THE YELLOW FROM A RAINBOW, PRISM, OR SPECTROSCOPE IS A PURE COLOR CONSISTING OF "YELLOW WAVELENGTHS."

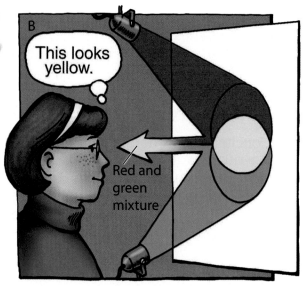

▶ B. THIS YELLOW IS PRODUCED DIFFERENTLY. IT COMES FROM THE YELLOW WE SEE WHEN RED AND GREEN LIGHTS ARE MIXED. NO WAVELENGTHS CORRESPONDING TO YELLOW LIGHT ARE REACHING THE EYE.

Red, Green, and Blue Entertainment

Although the visible spectrum contains a continuous range of colors from red to violet, our eyes can be tricked into seeing all these colors by mixing only three of them: red, green, and blue. In fact, we allow TV to use this trick every time we watch it.

Turn on a color TV. Look closely at the screen. If you have a magnifying glass, use it to study the picture on the screen. What do you see? If you look very closely at the screen, you will find that the picture the TV produces on the screen is made up of thousands of dots. Each dot is a red, green, or blue light. Where do these dots come from? How do they light up? How do they make a moving color picture?

Here's how it works in a traditional cathode ray TV (CRT). The inside of the front of the CRT screen is made up of thousands of dots, or pixels, arranged in groups of three. Each group contains one red, one green, and one blue pixel. Three beams of electrons (one for making each color) race across the screen targeting certain pixels through thousands of tiny holes positioned just behind the screen (in what is called the shadow mask). If a beam hits a pixel, the pixel glows. The beams use these pixels to draw a different picture about 30 times every second! As these pictures flash on the screen, they appear to our eyes as a moving image.

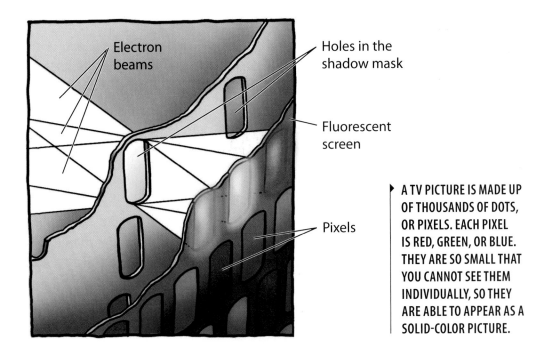

Electron beams

Holes in the shadow mask

Fluorescent screen

Pixels

▶ A TV PICTURE IS MADE UP OF THOUSANDS OF DOTS, OR PIXELS. EACH PIXEL IS RED, GREEN, OR BLUE. THEY ARE SO SMALL THAT YOU CANNOT SEE THEM INDIVIDUALLY, SO THEY ARE ABLE TO APPEAR AS A SOLID-COLOR PICTURE.

READING SELECTION
EXTENDING YOUR KNOWLEDGE

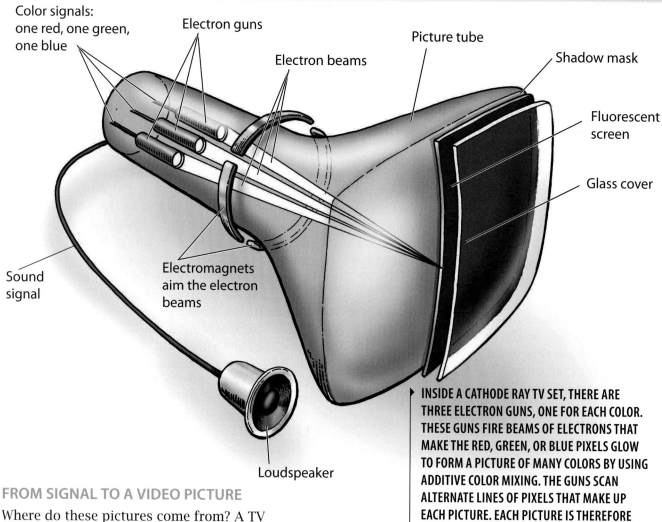

Color signals:
one red, one green,
one blue

Electron guns

Electron beams

Picture tube

Shadow mask

Fluorescent
screen

Glass cover

Electromagnets
aim the electron
beams

Sound
signal

Loudspeaker

FROM SIGNAL TO A VIDEO PICTURE

Where do these pictures come from? A TV receives signals, either as radio-wave signals through its antenna or satellite receiver or as signals via cable. It converts the signals to sound (audio) and pictures (video). The video part of the signal is further decoded into signals for red, green, and blue.

In a cathode ray TV, these signals are used to control the electron beams that light up the appropriate colored pixels on the screen. By lighting different pixels in a group, the beams trick the eye into seeing different colors—through additive color mixing. If the red and green pixels in one group are lit, your eyes see yellow. If the beam lights up all three in the group (red, green, and blue), your eyes see white. (This is the same type of color mixing

INSIDE A CATHODE RAY TV SET, THERE ARE THREE ELECTRON GUNS, ONE FOR EACH COLOR. THESE GUNS FIRE BEAMS OF ELECTRONS THAT MAKE THE RED, GREEN, OR BLUE PIXELS GLOW TO FORM A PICTURE OF MANY COLORS BY USING ADDITIVE COLOR MIXING. THE GUNS SCAN ALTERNATE LINES OF PIXELS THAT MAKE UP EACH PICTURE. EACH PICTURE IS THEREFORE SCANNED ONTO THE SCREEN TWICE—60 SCANS OR 30 COMPLETE PICTURES EACH SECOND.

you observed when you shined the flashlights through the colored filters.) By changing the combination and brightness of each of the colored pixels, the TV can create thousands of colors.

FINE-TUNING THROUGH TECHNOLOGY

The cathode ray TV is pretty heavy and large, and it consumes considerable power. Making it smaller and more energy-efficient, as well as improving the screen image, have led to the development of televisions that employ either plasma or LCD technologies.

DISCUSSION QUESTIONS

1. The reading selection says that TV can "trick" your eye into seeing different colors. How?

2. Many TVs allow you to adjust the color. How do you think that process works?

▶ IMAGINE WATCHING A TV LIKE THIS ONE. THE PICTURE IS ONLY IN BLACK AND WHITE. WHY DO YOU THINK IT WAS EASIER AND CHEAPER TO MAKE BLACK-AND-WHITE TV SETS?

PHOTO: Courtesy of the National Library of Medicine

A plasma TV is a high definition (HDTV) alternative to the traditional cathode ray televisions. In plasma TV panels, the back of each cell is coated with a phosphor. The ultraviolet photons emitted by the plasma excite these phosphors to give off colored light. The operation of each cell is much like that of a fluorescent lamp. A plasma TV provides sharp images and vibrant colors, especially when used with high definition broadcasts. The display panel, or viewing screen, is large, but only about 6 cm (2.4 inches) thick, while the total thickness, including electronics, is less than 10 cm (4 inches). Plasma screens, however, are made from glass, which reflects more light and weighs a lot. The reflections off the glass tend to create a glare although there continue to be technological advancements in this area.

LCD televisions display pictures using a liquid crystal display. One very important trait of liquid crystals is that they're affected by electric current. One type of liquid crystal, called twisted nematics (TN), is naturally twisted. Applying an electric current to these liquid crystals will untwist them to varying degrees, depending on the current's voltage. LCDs use these liquid crystals because they react predictably to electric current in such a way as to control the passage of light. The manipulation of the current allows for the variation in the image displayed. LCD technology makes for thinner and much lighter televisions. They also require much less power to operate than cathode ray televisions.

The technology of television has come a long way since its inception. What improvements can we look forward to in the future? ∎

HOW IS LIGHT REFLECTED?

WHAT IS THIS NAVAL OFFICER LOOKING THROUGH? WHAT IS IT USED FOR? HOW DOES IT WORK?

PHOTO: National Archives and Records Administration

INTRODUCTION

In Lessons 5 and 8 you observed what happens to light as it passes through a triangular prism. What happens when light strikes an object that it cannot pass through? In this lesson, you will conduct an inquiry on what happens to light when it strikes a plane mirror. You then will apply what you have discovered to help you to predict and control the direction of a ray of light.

OBJECTIVES FOR THIS LESSON

Observe and measure reflection of light rays off a plane mirror.

Use mirrors to redirect light rays.

MATERIALS FOR LESSON 10

For your group and another group to share

1	ray box
1	ray box lid
1	60-W clear halogen lightbulb
1	extension cord
1	bulb holder
2	narrow-slit ray box masks
2	wide-slit ray box masks
2	no-slit ray box masks

For your group

1	white screen
1	comb
1	large mirror
2	small mirrors
1	sheet of white paper
1	box of colored pencils
1	protractor
4	plastic stands
1	copy of Inquiry Master 10.1: Protractor Paper for Inquiry 10.1

GETTING STARTED

1 Have one member of your group obtain the materials.

2 Work with your group to set up the ray box (as shown in Figure 10.1) so that it produces a single ray of light.

3 Experiment with placing the large mirror into the path of the ray of light. Record your observations in your science notebook. 🗗

SAFETY TIP

Do not touch the lightbulb. It gets hot and may burn your fingers.

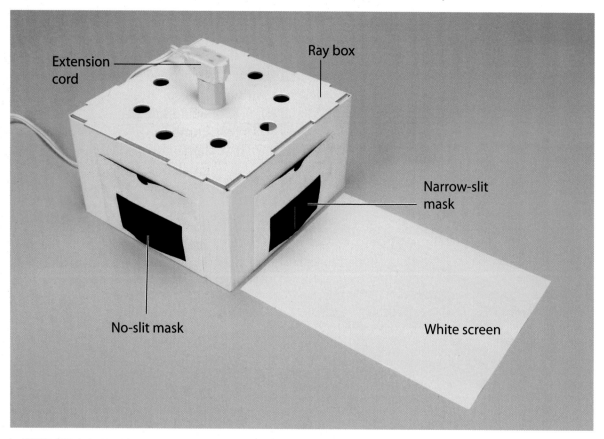

Extension cord
Ray box
Narrow-slit mask
No-slit mask
White screen

▶ SET UP THE RAY BOX SO THAT IT PRODUCES A SINGLE RAY OF LIGHT.
FIGURE **10.1**
PHOTO: ©2009 Carolina Biological Supply Company

MEASURING REFLECTION

PROCEDURE

1 Have one member of your group use the scissors to cut along the cutout lines on Inquiry Master 10.1: Protractor Paper for Inquiry 10.1. Remember to record all your responses in your science notebook. 📄

2 Working with your group, set up the apparatus as shown in Figure 10.2. The base of the large mirror should lie along the baseline of the protractor paper. Position the mirror so that its center approximately matches the center of the protractor's baseline. Throughout this inquiry, the mirror should stay in this position on the protractor paper. (If you prefer, you may tape your protractor paper to your mirror in the correct position.)

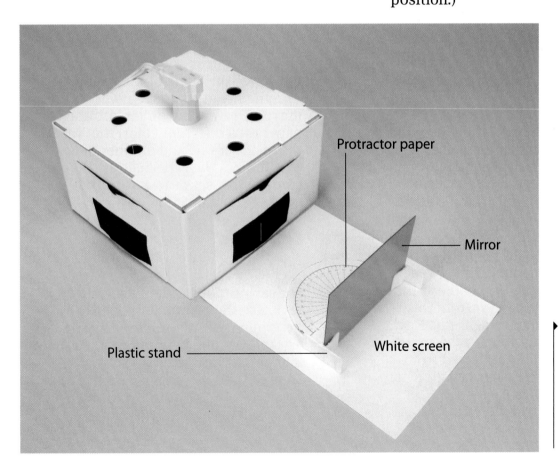

Protractor paper

Mirror

Plastic stand

White screen

▶ **POSITION THE LARGE MIRROR, WHITE SCREEN, AND PROTRACTOR PAPER AS SHOWN.**
FIGURE **10.2**
PHOTO: ©2009 Carolina Biological Supply Company

Inquiry 10.1 continued

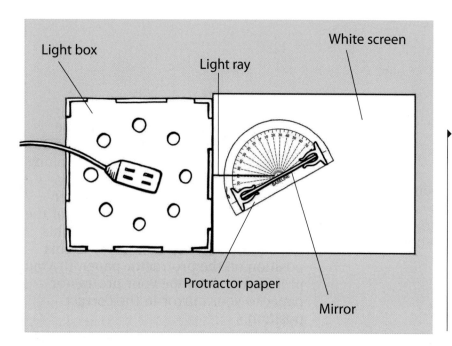

MOVE THE MIRROR AND PROTRACTOR PAPER SO THAT THE RAY OF LIGHT FROM THE NARROW SLIT PASSES DOWN THE 60° LINE OF THE PROTRACTOR PAPER. THE RAY SHOULD STRIKE THE MIRROR WHERE THE 0° LINE (THE LINE PERPENDICULAR TO THE MIRROR, WHICH IS CALLED THE NORMAL) MEETS AT THE CENTER OF THE BASELINE.

FIGURE **10.3**

3 Move the mirror and protractor paper so that the ray of light from the narrow slit passes down the 60° line of the protractor paper. The ray should strike the mirror where the 0° line (the line perpendicular to the mirror, which is called the normal) meets at the center of the baseline, as shown in Figure 10.3.

4 On the protractor paper, use a colored pencil and a ruler to draw a line that accurately follows the ray of light from the center of the ray box to the large mirror. This ray is called the incident ray. Next, draw a line that follows the center of the ray of light reflected from the mirror. This ray is called the reflected ray.

5 Move the mirror and protractor paper to change the angle of the incident ray. Aim the ray at the same point on the mirror. Draw a line on the protractor paper (in a different color) that follows the incident ray, and another line that follows the reflected ray. Do not confuse this set of rays with the rays you recorded in Step 4.

6 Repeat Step 5 a few more times.

7 For each set of incident and reflected rays, use the protractor paper to determine the angle between each ray and the line perpendicular to the mirror (the line labeled "Normal" in Figure 10.4).

8 Design and then draw a table where you can record your data in your science notebook. 🖉

9 Compare your results with those from another group. What can you conclude from your results? Record your ideas.

10 Move the white screen and the mirror to the side of the ray box with the wide-slit mask. Use the comb against the wide-slit mask (as shown in Figure 10.5) so that it produces multiple rays. Direct the rays at the mirror. Draw what you observe. Be prepared to share your observations with the class.

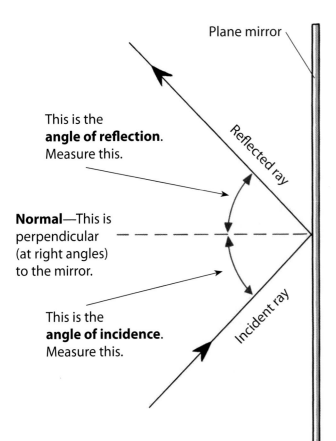

Plane mirror

This is the **angle of reflection**. Measure this.

Reflected ray

Normal—This is perpendicular (at right angles) to the mirror.

This is the **angle of incidence**. Measure this.

Incident ray

▸ USE THIS DIAGRAM TO IDENTIFY THE ANGLE OF INCIDENCE AND ANGLE OF REFLECTION FOR EACH RAY. USE THE PROTRACTOR PAPER TO MEASURE THESE ANGLES.
FIGURE **10.4**

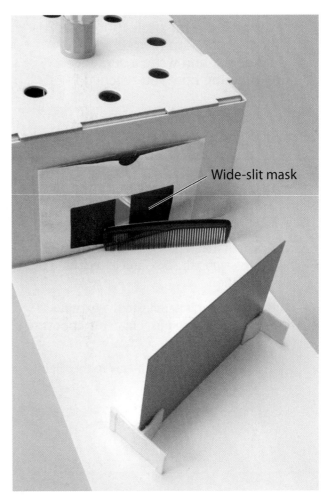

Wide-slit mask

▸ DIRECT MULTIPLE RAYS AT THE MIRROR. WHAT DO YOU OBSERVE?
FIGURE **10.5**
PHOTO: ©2009 Carolina Biological Supply Company

INQUIRY 10.2

CHANGING THE PATH OF A LIGHT RAY

PROCEDURE

1. Look at the ray box setup from Inquiry 10.1. Predict the angle at which the mirror will have to be held so that the ray of light turns through a right angle (90°). Record your prediction in your science notebook and write a sentence explaining how you reached it. ☞

2. Test your prediction. Use the protractor to measure the angles.

3. Keep your apparatus in place, and use a second mirror to redirect the light ray so that it continues on each of the following paths:

   A. a path that is parallel to the existing ray leaving the box, and in the same direction

   B. a path that is parallel to the existing ray leaving the box, but in the opposite direction

   C. a path that returns the ray to the slit in the ray box

READING SELECTION

BUILDING YOUR UNDERSTANDING

REDIRECTING LIGHT, IMAGES, AND THE LAW OF REFLECTION

You have determined that the angle of reflection from a plane mirror equals the angle of incidence. This rule is called the law of reflection. By altering the angle of a mirror with respect to the incident ray, you can alter the direction in which the ray travels. By reflecting light from one mirror to another, light can be made to change direction many times and can be directed around opaque objects. ∎

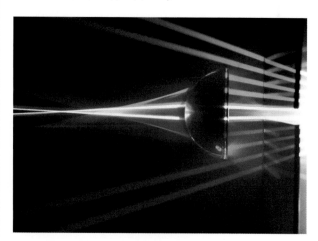

▶ **LIGHT CAN BE REFLECTED FROM ONE MIRRORED SURFACE TO ANOTHER.**

PHOTO: Susan Schwartzenberg, © Exploratorium, www.exploratorium.edu

4 Which group in the class can construct the most complicated ray path using multiple mirrors? Use all the mirrors and plastic stands at your disposal. Spend a few minutes constructing your ray path. What problems do you encounter? Can you reflect light from the same ray off the same mirror more than once? Compare your ray path with those constructed by other groups.

SAFETY TIP

Turn off the ray box immediately after you finish using it. Allow the ray box to cool.

5 Read "Redirecting Light, Images, and the Law of Reflection."

 REFLECTING
ON WHAT
YOU'VE DONE

1 Look back at what you recorded on Student Sheet 1 for Inquiry 1.4. Can you explain now how you were able to use a mirror to look behind you? Redraw the diagram in your science notebook. Add arrows to the rays to indicate the direction the light is traveling, and label the rays and angles.

2 Write a few sentences, using the correct terminology, to describe what is happening in the diagram you drew.

Abu Ali Hasan Ibn al-Haytham

Just like art, literature, and music, science has its roots in many cultures. At the turn of the first millennium—1000 A.D.—the culture and learning of the Greeks, although still important, were ancient history. The Roman Empire's rule over Europe had long before fallen apart, and much of Europe was just emerging from the so-called Dark Ages. But other cultures were blossoming.

Islamic influence was expanding, and the Islamic world stretched from the borders of India to North Africa and parts of southern Europe. Islamic scholars were considered the most learned in the world. They worked in all aspects of art, mathematics, and science, and had access to the world's best universities and libraries. They were well educated and widely traveled.

One such scholar was an Arab born in Basra, an important city located in what is now Iraq. His name was Abu Ali Hasan Ibn al-Haytham. He is better known in the Western world by a shortened version of his name, Alhacen. Alhacen was interested in many subjects, but he specialized in mathematics, astronomy, and physics. One of his favorite subjects was optics.

Alhacen liked to design inquiries to test his theories, as well as those of others. He studied light from sources such as lamps, fire, the Moon, and the Sun. He suggested that light was a single phenomenon, regardless of its source or color. He conducted experiments to determine that light travels in straight lines. He also studied what happens to light when it enters a transparent material. He used a prism to make a colored spectrum. He even correctly suggested the cause of twilight.

▶ MANY CULTURES HAVE CONTRIBUTED TO CURRENT SCIENTIFIC KNOWLEDGE. ALHACEN, OR MORE CORRECTLY, ABU ALI HASAN IBN AL-HAYTHAM, AN ARAB FROM THE ANCIENT CITY OF BASRA, IS ONE OF THE FOUNDERS OF THE SCIENCE OF OPTICS. HE ALSO WAS ONE OF THE EARLIEST DEVELOPERS OF THE SCIENTIFIC METHOD.

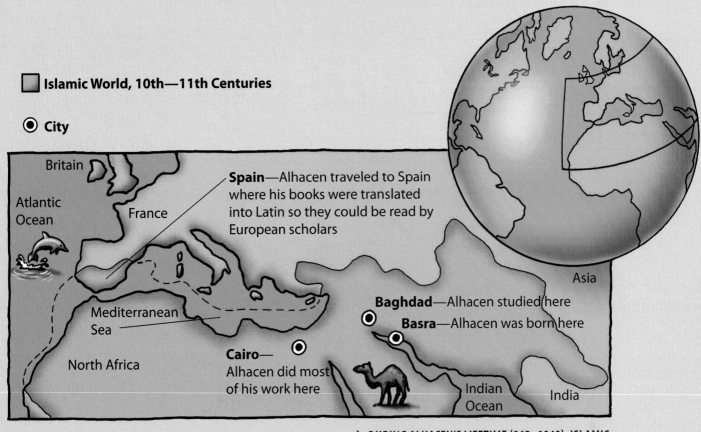

Islamic World, 10th—11th Centuries

City

Britain

Atlantic Ocean

France

Spain—Alhacen traveled to Spain where his books were translated into Latin so they could be read by European scholars

Mediterranean Sea

North Africa

Cairo—Alhacen did most of his work here

Baghdad—Alhacen studied here

Basra—Alhacen was born here

Asia

Indian Ocean

India

▶ DURING ALHACEN'S LIFETIME (965–1040), ISLAMIC RULE AND SCHOLARSHIP STRETCHED FROM THE INDIAN OCEAN TO THE NORTH ATLANTIC.

Alhacen investigated how the eyes worked. He was probably the first person to record the idea that light travels from an object to the eye, and not in the opposite direction.

Although Alhacen studied human vision, he is best known for his research on reflection. Alhacen's studies of reflection confirmed earlier Greek theories about reflection from plane mirrors—including the law of reflection, which states that the angle of incidence equals the angle of reflection. He then applied these laws to curved mirrors.

READING SELECTION

EXTENDING YOUR KNOWLEDGE

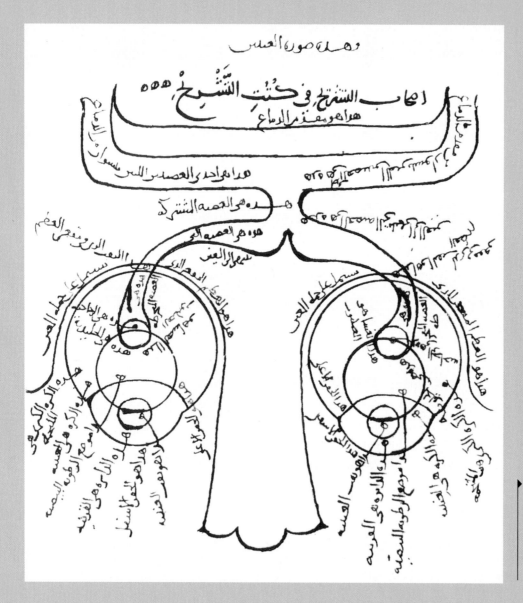

▶ **ALHACEN INVESTIGATED HOW THE EYES WORKED. HERE IS ONE OF HIS DIAGRAMS OF THE HUMAN VISUAL SYSTEM.**

PHOTO: Private donor

Alhacen wrote about 200 books. He developed a scientific methodology, an experimental approach to explaining the natural world. This approach was adopted 500 years later by European scientists at the forefront of a new revolution in the arts and sciences known as the Renaissance. ∎

▶ WHY CAN YOU SEE SUNLIGHT WHEN THE SUN IS BELOW THE HORIZON AND NOT VISIBLE IN THE SKY? ALHACEN SUGGESTED THAT LIGHT FROM THE SUN BELOW THE HORIZON TRAVELED AROUND EARTH BY SOMEHOW BEING REFLECTED FROM HIGH IN THE ATMOSPHERE. WE NOW KNOW THIS PROCESS INVOLVES LIGHT BEING SCATTERED FROM MOLECULES HIGH IN THE ATMOSPHERE. THIS SCATTERING ALSO BENDS LIGHT AROUND EARTH DURING A LUNAR ECLIPSE AND EXPLAINS WHY THE MOON OFTEN REFLECTS A DIM RED LIGHT DURING A TOTAL ECLIPSE.

PHOTO: Commander John Bortniak, NOAA Corps

 DISCUSSION QUESTIONS

1. Alhacen was one of the early founders of the "experimental approach." What does this mean?

2. Choose one of Alhacen's theories mentioned in the reading selection. What experiments would you perform to test that theory?

INTRODUCING REFRACTION

INTRODUCTION

Have you looked through a window today? For that matter, did you observe your hands while washing them this morning, or look at juice through the side of a glass during breakfast? If you did, you observed light passing from one transparent material to another. What happens when light strikes a transparent object or material like a window or the surface of water? You may already have some ideas.

Here are a few questions to ask yourself. Does all the light striking a transparent object enter it? Does the light entering a transparent object travel in a straight line through it? Does the light entering a transparent object leave it? Does the composition or shape of a transparent object affect the behavior of light? Get ready to share your ideas before you launch into three inquiries that will help test your ideas—and raise other questions—about how light interacts with transparent objects.

▶ WE LOOK THROUGH TRANSPARENT MATERIALS—SOLIDS, LIQUIDS, OR GASES—ALL THE TIME. DO TRANSPARENT MATERIALS AFFECT THE APPEARANCE OR POSITION OF WHAT WE OBSERVE?

PHOTO: skozan/creativecommons.org

OBJECTIVES FOR THIS LESSON

- Discuss how light interacts with transparent objects.

- Make observations through a transparent block.

- Observe and make measurements of a light ray as it interacts with a transparent block.

- Use standardized terms to describe the behavior of a light ray as it interacts with a transparent block.

▶ **MATERIALS FOR LESSON 11**

For you

1	copy of Student Sheet 11.2: Shining Light Into a Transparent Block
1	copy of Student Sheet 11.3: Measuring Refraction in a Transparent Block

For your group

1	transparent block
1	white screen
1	copy of Inquiry Master 11.1: Protractor Paper for Inquiries 11.2 and 11.3
1	protractor
1	metric ruler, 30 cm (12 in)
1	box of colored pencils
1	pair of scissors

For your group and another group to share

1	ray box
1	ray box lid
1	60-W clear halogen lightbulb
1	extension cord
1	bulb holder
2	narrow-slit ray box masks
2	no-slit ray box masks

GETTING STARTED

1 Listen as a classmate reads the Introduction aloud. Use the questions in the Introduction to help you with your thinking in Steps 2 and 3.

2 Discuss with your group where, in previous lessons, you have observed light interacting with transparent objects. Discuss what you observed in each case.

3 With the class, brainstorm examples of these and other observations you have made of light interacting with transparent objects and materials.

INQUIRY 11.1

LOOKING THROUGH A TRANSPARENT BLOCK

PROCEDURE

1 One member of your group should obtain the materials as directed by your teacher. There is no student sheet for this inquiry. Record your responses in your science notebook. ☞

2 Examine the transparent block. Take turns looking through the block from different angles. Discuss what you observe with your group. Make a list of all your observations.

3 Rest the block on top of the page of this book. Record what you observe.

4 Take turns looking sideways through the block. Hold your finger up behind the block. Look at your finger through the block and slowly rotate the block from side to side (see Figure 11.1). What do you observe when you rotate the block? Record your observations and share them with your partners.

5 What do you think is happening? Be prepared to share your observations and ideas with the class.

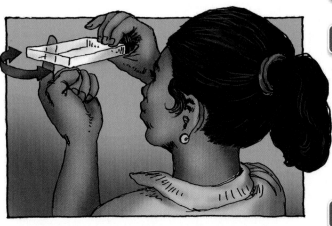

▶ **LOOK AT YOUR FINGER THROUGH THE BLOCK. WHAT DO YOU OBSERVE WHEN YOU ROTATE THE BLOCK?**
FIGURE **11.1**

INQUIRY 11.2

SHINING LIGHT INTO A TRANSPARENT BLOCK

PROCEDURE

1 Set up the ray box and white screen so that the side of the ray box at which your group is working will produce a single ray of light.

2 Use the scissors to cut along the cutout lines on Inquiry Master 11.1: Protractor Paper for Inquiries 11.2 and 11.3.

3 Place the protractor paper on top of the white screen. The curved side of the protractor diagram should face the mask (see Figure 11.2). It should be no more than 3 centimeters from the ray box.

4 Place the transparent block on the paper so that it is parallel to the ray box and its top edge is along the baseline of the protractor paper (see Figure 11.2).

5 Plug in the ray box. Make sure the ray passes through your block.

SAFETY TIP

Do not touch the lightbulb. It gets hot and may burn your fingers.

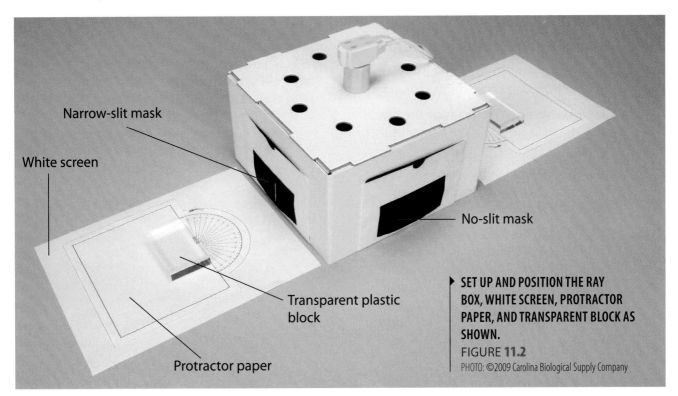

Narrow-slit mask

White screen

No-slit mask

Transparent plastic block

Protractor paper

▶ SET UP AND POSITION THE RAY BOX, WHITE SCREEN, PROTRACTOR PAPER, AND TRANSPARENT BLOCK AS SHOWN.
FIGURE **11.2**
PHOTO: ©2009 Carolina Biological Supply Company

Inquiry 11.2 continued

6 Shine the ray along the 0° line (the normal) of the protractor paper. Observe what happens when the light ray strikes the transparent block.

A. On Student Sheet 11.2: Shining Light Into a Transparent Block, complete the diagram in Step 1 by recording what you observe. Draw all the rays you observe. Do not forget to draw arrows to show the direction you think the light is traveling.

B. Describe what happens to the ray of light as it strikes the block.

7 Very slowly turn the block clockwise on the protractor paper (as shown in Figure 11.3). Carefully observe the behavior of the light as it strikes the rotating block. Use the following to guide your observations:

A. As you rotate the block, stop about every 15–20° and use a diagram and words to record what you observe in Step 3 on the student sheet.

B. What do you notice about the angle of the reflected ray as you rotate the block? Record your observations under Step 4.

C. In Step 5, describe how your observations fit with what you know about reflection.

D. Can you observe the ray inside the block? Describe it in Step 6 on the student sheet.

E. What do you observe about the position and direction of the ray as it leaves the block? Write a description under Step 7 on Student Sheet 11.2.

F. What do you think happens to the direction of the light ray as it enters and leaves the block? Record your ideas under Step 8.

READING SELECTION

BUILDING YOUR UNDERSTANDING

REFLECTION AND REFRACTION

Light changes direction when it reflects off a surface. We call this change in direction reflection. Light also may change direction when it travels from one transparent material into another. (What were the two transparent materials involved in your investigation?) This change in the direction of light is called refraction. Once light has been refracted, it continues in a straight line within that material until it strikes another surface. ■

8 Read "Reflection and Refraction."

9 Using your observations and the information from the reading selection, write an explanation of where the refraction of the ray took place in this inquiry on your student sheet.

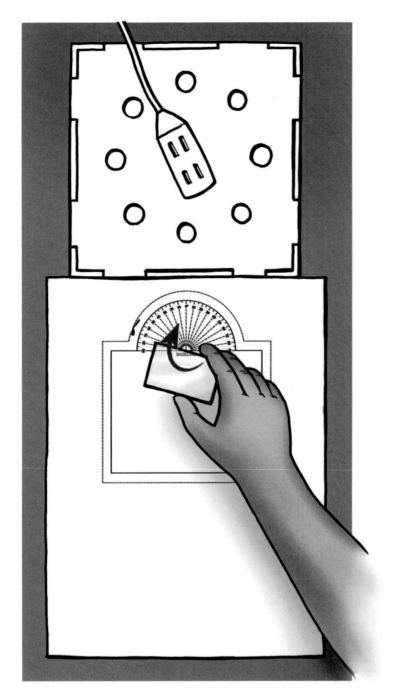

▶ **VERY SLOWLY ROTATE THE BLOCK CLOCKWISE.**
FIGURE **11.3**

INQUIRY 11.3

MEASURING REFRACTION IN A TRANSPARENT BLOCK

PROCEDURE

1 Reposition the block and protractor paper so that the light ray strikes the block along the 0° line on the protractor paper.

2 On the protractor paper, use a colored pencil to draw around the block (see Figure 11.4). Draw lines showing where light enters and leaves the block.

3 Turn both the transparent block and protractor paper 20° clockwise (see Figure 11.5). (The front face of the block should still lie along the baseline of the protractor paper.) The light ray now has an angle of incidence of 20°.

4 Use a different colored pencil to draw the rays entering and leaving the block as before.

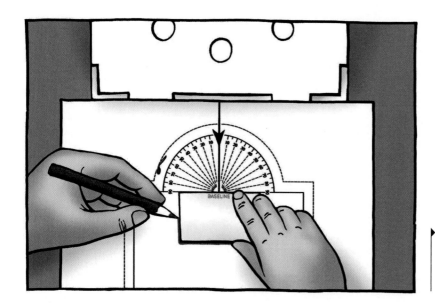

▶ DRAW AROUND THE BLOCK. DRAW LINES SHOWING WHERE LIGHT ENTERS AND LEAVES THE BLOCK.
FIGURE **11.4**

5 Rotate the block and protractor paper 20° more. Use a third colored pencil to draw the rays.

6 Remove the block from the paper. For each different angle of the incident ray, use the ruler and appropriate colored pencil to draw in the path of the light rays as they pass through the block.

7 Transfer what you have recorded onto the three diagrams of blocks in Step 1 on Student Sheet 11.3: Measuring Refraction in a Transparent Block. (You may find it helpful to rotate the protractor paper so that the angle of the block's outline matches the angle shown in the diagram.)

ROTATE THE TRANSPARENT BLOCK AND THE PROTRACTOR PAPER 20° CLOCKWISE. USE A DIFFERENT COLORED PENCIL TO DRAW RAYS ENTERING AND LEAVING THE BLOCK.
FIGURE **11.5**

Inquiry 11.3 continued

8 Compare how light behaved when the block was in the three different positions. Did the light ray entering the block always behave in the same way? Record your ideas.

9 Complete Steps 3 through 6 on Student Sheet 11.3. Look at the diagrams you have completed for the blocks with angles of incidence of 20° and 40°.

- Describe exactly what happens to the light ray as it passes from air into the transparent block at each angle.

- Describe exactly what happens to the light ray as it passes from the transparent block into the air.

- How does changing the angle at which the light ray strikes the block affect the position of the ray as it leaves the block?

- Is there any relationship between the direction of the ray striking the block and the direction of the ray leaving the block?

10 Read "Introducing Refraction."

11 On one of the diagrams you completed on Student Sheet 11.3, identify the incident, reflected, refracted, and emergent rays. Label them on your diagram.

12 When light passes into, through, and out of the transparent block, where does refraction occur? Record your ideas.

13 When light passes from air into transparent plastic, does it behave the same way as it does when it passes from transparent plastic into the air? Describe any differences or similarities that occur. Record your description.

SAFETY TIP

Turn off the ray box immediately after you finish using it. Allow the ray box to cool.

INTRODUCING REFRACTION

When a ray of light (the incident ray) strikes the transparent block, some light is reflected (the reflected ray) and some light enters and passes through the block. If the ray enters at a right angle (along the normal) to the block's surface, the light passes through the block in a straight line. If the ray has a different angle of incidence, the ray bends at the boundary—it is refracted—between the air and the transparent block. The light ray inside the block is called the refracted ray. The angle between the normal and the refracted ray is called the angle of refraction.

The refracted ray travels in a straight line and emerges from the block. At this point, the ray emerges from the block and changes direction again because it has been refracted by passing through the second surface of the block to re-enter the air. This refracted ray is called the emergent ray. Look at the diagram—it shows some of the rays you may have seen entering and leaving the transparent block. ■

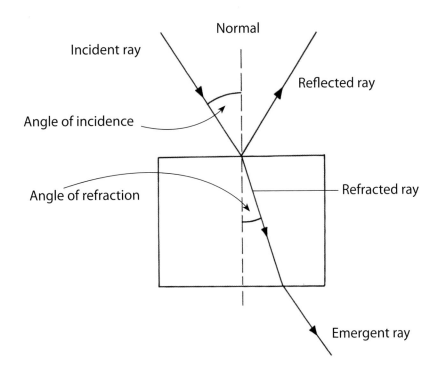

Inquiry 11.3 continued

14 Write in values for the angles of incidence on each diagram.

15 Use a protractor to measure the angles (on the protractor paper) of refraction you obtained. Write these on each diagram.

16 Write a few sentences to describe the effect that changing the angle of the incident ray has on the angle of refraction.

REFLECTING
ON WHAT
YOU'VE DONE

1 Record your response to the following in yo science notebook. Be prepared to share yo responses with the class.

A. What is the differen between reflection and refraction?

B. Write a paragraph summarizing what you have learned about the behavior of light after it strik and passes throug a transparent block When writing your description, try to u the correct names for the rays and the angles they have to the normal.

2 Review the question bank generated in Lesson 1. Can you answer any more questions now? Ident those that you feel comfortable answerin

3 Read "Refractive Inde> and Wet Pants."

Refractive Index
AND WET PANTS

PHEW!
It's a scorching day, and that creek looks very inviting to a hot, tired hiker. His feet are aching, and he's dying to slip off his stiff hiking boots and thick socks and wade in that nice cool water. In he goes, and . . . Splash! Instead of being up to his ankles, the water is up to his knees. Why was the water much deeper than it appeared—and how will the hiker dry his pants? If only he had paid attention in science class! Why? Because he would have known that the apparent depth of the water was an illusion. Look at the picture of the hiker. Can you see how the hiker's brain was tricked by refraction?

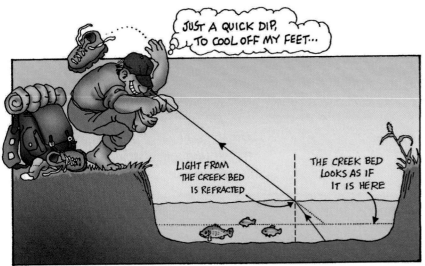

THE WATER IS DEEPER THAN IT LOOKS. KNOWLEDGE OF HOW LIGHT IS REFRACTED WHEN IT PASSES FROM WATER INTO THE AIR MIGHT HAVE SAVED THIS HIKER THE EMBARRASSMENT OF WET PANTS.

READING SELECTION

EXTENDING YOUR KNOWLEDGE

► **REFRACTION BETWEEN THE AIR AND WATER MAKES THIS STUDENT LOOK DISTORTED. WHAT CAUSES REFRACTION?**

PHOTO: Eric Long, Smithsonian Institution

WHAT CAUSES REFRACTION?

Light is refracted when it changes speed. Light changes speed when it passes from one transparent material into another. For example, light travels slower in water than in air—about three-quarters as fast. Light therefore bends when it passes from air into water. (Do you remember the nail in water in Lesson 1? Did it appear bent?) The difference between the speed of light in two transparent materials determines how much the light bends as it passes between them. The bigger the difference in the speed of light in the two materials, the more bending, or refraction, takes place when the light passes from one material into another.

REFRACTIVE INDEX

Scientists find it useful to compare the light-bending abilities of different transparent materials. They make this comparison using something called the refractive index. The refractive index of a transparent material is defined as the speed of light in a vacuum divided by the speed of light in the transparent material. Each transparent material has a different refractive index.

The slower the speed of light in a material, the higher its refractive index. In a vacuum—where there is no matter to slow down the light—light travels at its fastest speed. The refractive index of a vacuum is 1. This is the lowest refractive index. In glass, light travels much slower, about two-thirds the speed of light in a vacuum. The refractive index for glass is about 1 divided by two-thirds—a refractive index of about 1.5. Table 1 shows refractive indices of a vacuum and of different transparent materials.

In which of these transparent materials does light travel the slowest?

TABLE 1 REFRACTIVE INDICES OF SOME TRANSPARENT MATERIALS

TRANSPARENT MATERIALS	REFRACTIVE INDEX
VACUUM (A VACUUM CONTAINS NO MATTER)	1.00
AIR	SLIGHTLY HIGHER THAN 1.00 (1.000293)
GLASS	1.53
DIAMOND	2.42
TRANSPARENT PLASTIC	1.50
WATER	1.33

READING SELECTION
EXTENDING YOUR KNOWLEDGE

▶ THE ARCHERFISH SQUIRTS WATER AT INSECTS TO KNOCK THEM OFF OVERHANGING WATERSIDE PLANTS. ARCHERFISH HAVE EVOLVED A SHOOTING TECHNIQUE TO DEAL WITH THE REFRACTION THAT TAKES PLACE BETWEEN WATER AND AIR. EACH TIME THE FISH SHOOTS A STREAM OF WATER IT TAKES INTO ACCOUNT THAT THE INSECTS ARE NEARER, AND AT A DIFFERENT ANGLE, THAN THEY APPEAR.

PHOTO: Courtesy of Shelby Temple

BIRD BRAINS AND FISHY PHYSICS

If you know the refractive index of two materials, you can predict which way light will bend when it passes from one to another. Light passing from a material with a lower refractive index to one with a higher refractive index bends toward the normal (just as you observed when light passed from the air into the transparent plastic block).

What happens when light travels in the opposite direction? When light passes from a material with a higher refractive index into one with a lower refractive index (for example, from the transparent plastic block into the air), it is refracted in the opposite direction. The light ray bends away from the normal.

If the hiker had known this, could he have avoided getting his pants wet? Look back at the picture of the soggy hiker. Light reflected from the creek bed travels from water into the air. Air has a lower refractive index than water. As light passed from the water into the air it was refracted away from the normal. This made the water look shallow. The hiker was tricked by refraction!

Some birds that are expert fishers, such as the great blue heron, don't make the mistake the hiker made. They take refraction into account when they lunge underwater for their prey. They must lunge at a position deeper and at an angle different than where the fish appears to be. Did you know that birdbrains were so good at physics?

Now imagine you're a fish looking up from the water into the air. You would have the reverse problem of the hiker or the heron. Light traveling from the air into the water is refracted in the opposite direction, toward the normal. From underwater, objects look farther away than they actually are.

Can you use refractive indices to predict how light will behave when it passes from water to glass? Will a ray of light be refracted toward or away from the normal? ■

DISCUSSION QUESTIONS

1. Rank the materials in Table 1 from fastest to slowest light speed when light passes through them.

2. When do we or could we make use of knowledge of refraction (besides deciding whether to jump into a stream)?

MODELING LIGHT

INTRODUCTION

You have spent the last few lessons investigating some of the characteristics of light. You looked at how light is produced and how it spreads out from a light source. You determined that light travels in straight lines and can travel through air and the vacuum of space. You compared the behavior of light when it strikes transparent, opaque, and translucent objects. You used what you have discovered to explain the formation of shadows. You already know quite a lot about how light behaves!

But what gives light these characteristics? How are the characteristics of light related to one another? What is the exact nature of light? Is there some way you can better understand what light is? Can you use your knowledge of the nature of light to explain why it behaves in particular ways? Can you predict how it will behave in different situations?

One way you can better understand what light is and why it behaves as it does is by using scientific models. After you revisit your ideas about the nature of light, you will discuss the nature of scientific models and then use two scientific models for light. At the end of this lesson, you will be asked to compare how well these models explain what you observed in previous lessons.

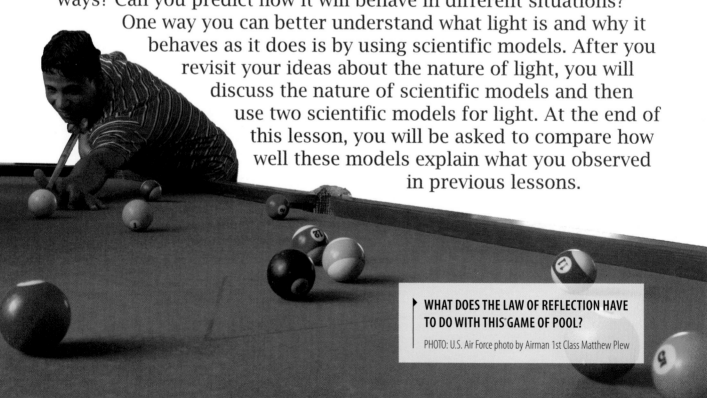

▶ **WHAT DOES THE LAW OF REFLECTION HAVE TO DO WITH THIS GAME OF POOL?**

PHOTO: U.S. Air Force photo by Airman 1st Class Matthew Plew

OBJECTIVES FOR THIS LESSON

- Discuss the nature of light.

- Use and compare scientific models for light.

▶ MATERIALS FOR LESSON 12

For you

1	copy of Student Sheet 12.1: Using Particles to Model Light
1	copy of Student Sheet 12.2: Using Waves to Model Light

For your group

15	ball bearings (in a resealable plastic bag)
1	straight metal barrier
1	transparent tray
1	cardboard tube
1	transparent cup
1	wooden dowel
1	flashlight
2	D-cell batteries
2	wooden blocks
1	dropping bottle
1	metric ruler, 30 cm (12 in)
1	cork
1	sheet of white paper
	Hardcover books from the classroom

GETTING STARTED

1. In Lesson 1, you were asked the question, What is light? Think about the question again. Discuss any new ideas you have with your group. Have your ideas about the nature of light changed? Have other members of your group reached similar conclusions? In your science notebook, write what you now think light is. 🖉

2. If light is a form of energy, how does it move from one place to another? Look at Figure 12.1. What is happening in each photograph? These pictures may give you some ideas about how energy can be moved from one place to another.

3. Discuss your ideas with the class.

4. Read "Scientific Models" on page 148.

▶ **HOW WAS ENERGY FROM A DISTANT EARTHQUAKE TRANSFERRED TO THESE BUILDINGS?**

PHOTO: National Geophysical Data Center

▶ **HOW IS ENERGY BEING MOVED FROM ONE PLACE TO ANOTHER?**
FIGURE **12.1**

WHAT MAKES THIS SURFER MOVE? WHERE DID THE
ENERGY COME FROM? HOW DID IT GET TO THE SURFER?

PHOTO: Michael "Mike" Baird, flickr.bairdphotos.com

WHOEVER KICKED THIS BALL IS ABOUT TO TRANSFER
THE ENERGY TO ANOTHER PLAYER'S HEAD. HOW DID
THIS ENERGY TRANSFER TAKE PLACE?

PHOTO: U.S. Army photo by Spc. Charles W. Gill

SCIENTIFIC MODELS

In this lesson, you will investigate two scientific models for light. A scientific model is a way of thinking about how something works. It is not a copy of an object, like a plastic model of an airplane or the human body. Scientific models help scientists understand complex processes or systems that are difficult to understand or observe.

You may already have used scientific models. For example, have you ever used a model of the structure of the atom? You might have also used a model to explain how electrical energy moves around a circuit.

Before scientists construct a scientific model, they closely observe what they are studying. Next, they try to link these observations. Then they design a model that behaves in the same way. The model may be a mathematical one. Today's scientists often simulate models on computers. Or they may make models from materials that behave like the thing they are modeling.

Scientists may use scientific models to help predict how the things they are studying will behave. A good model can be used to make accurate predictions. For example, weather forecasters use computer models of Earth's weather (based on lots of data collected from around the world).

Scientists commonly use two scientific models for light: the particle (or photon) model and the wave model. Both can be partly represented using materials that are easy to observe and that sometimes behave in ways similar to light. You will use both models in this lesson.

Accurate models behave just like the things they are modeling. But most scientific models have limitations. They can demonstrate and help explain only some of the observations made by scientists. You will compare the particle and wave models with your own observations and knowledge about light. Is one model better? Use what you have learned about light to think about and evaluate these two models. ■

INQUIRY 12.1

USING PARTICLES TO MODEL LIGHT

PROCEDURE

1 One member of your group should obtain the materials. You will record your observations and explanations for this inquiry on Student Sheet 12.1: Using Particles to Model Light.

2 In this inquiry, you will use small ball bearings to model the behavior of light. Pick up the plastic bag of ball bearings. Look at and handle them. Discuss with your group how you could use a ball bearing to transfer energy from one place to another.

3 To model how light travels, roll a few ball bearings down the tube and across the flat surface of the transparent tray (see Figure 12.2). Record any energy transformations taking place as the ball bearings move from the top of the tube until after they strike the end of the tray.

4 How do the ball bearings behave as they travel across the flat surface? Write a description of how this aspect of the ball bearing model compares with what you know about light.

▶ ROLL A FEW BALL BEARINGS DOWN THE TUBE AND ACROSS THE TRANSPARENT TRAY.
FIGURE **12.2**

Inquiry 12.1 continued

▶ HOLDING THE END OF THE TUBE ABOUT 1 CM ABOVE THE TRAY,
DROP ALL THE BALL BEARINGS INTO THE TUBE AT ONCE.
FIGURE **12.3**

5 How do you think light spreads out from a source? To model this, hold the tube vertically with one end about 1 cm above the surface of the tray. Use a small piece of scrap paper to help you drop all the ball bearings into the tube at once (as shown in Figure 12.3). Record a description of how the ball bearings behave. How do you think light spreads based upon what you see in this model?

6 To model reflection, place the metal barrier in the tray as shown in Figure 12.4. Tilt the cardboard tube to an angle of about 30°. Roll one ball bearing down the tube so that it strikes the barrier (see Figure 12.4).

7 Discuss with your group how the simple activity you have just done could be adapted to investigate whether the ball bearings follow the law of reflection.

8 Using the materials provided, devise a procedure to model the reflection of light from a plane mirror. Discuss the type and quantity of data you need to collect. Devise a table for your data.

9 Outline your procedure on Student Sheet 12.1. Use your procedure to collect data. Record your data and observations.

10 Discuss your results with your group. Record any conclusions/comments you may have about the model. Be prepared to share your procedures, observations, and data with the class.

ROLL THE BALL BEARING
DOWN THE CARDBOARD
TUBE WHILE HOLDING
THE TUBE AT A 30°
ANGLE.
FIGURE **12.4**

11 Compare what happened in your model with how light behaves. Record your ideas.

12 Discuss the following questions with your group:

A. Is the ball bearing model (usually called a particle model) for light a useful model?

B. What are its limitations? (How does it fall short in explaining what you already know about light?)

C. How could the model be improved?

13 Return the ball bearings to the resealable bag. Be sure to seal the top of the bag.

INQUIRY 12.2

USING WAVES TO MODEL LIGHT

PROCEDURE

1 Record your observations and explanations for this inquiry on Student Sheet 12.2: Using Waves to Model Light. Your teacher will show you how to set up and use a ripple tank.

2 Set up your ripple tank, using Figures 12.5 and 12.6 to guide you.

3 Float the cork in the water at one end of the tray. How can you transfer energy from your finger at one corner of the tray to the cork at the other end of the tray without touching the cork? Try out your ideas.

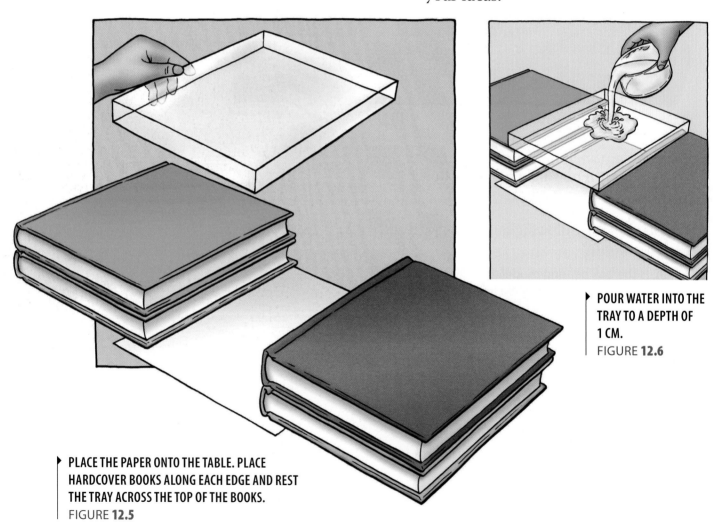

PLACE THE PAPER ONTO THE TABLE. PLACE HARDCOVER BOOKS ALONG EACH EDGE AND REST THE TRAY ACROSS THE TOP OF THE BOOKS.
FIGURE **12.5**

POUR WATER INTO THE TRAY TO A DEPTH OF 1 CM.
FIGURE **12.6**

▶ **HOLD THE FLASHLIGHT OVER THE
TRAY SO THAT THE LIGHT SHINES
THROUGH THE WATER.**
FIGURE **12.7**

4 Record what you did and what
you observed. Identify any energy
transformations that took place. List
them in Step 2 on Student Sheet 12.2.

5 Remove the cork from the tray. Turn
on the flashlight and hold it over the
center of the tray so that the light
shines though the water as shown in
Figure 12.7.

Inquiry 12.2 continued

6 To model how light travels, use the dropping bottle to drop one drop of water into the center of the tray. Look carefully at the paper.

    A. On Student Sheet 12.2, under Step 3, draw and describe what you observe.

    B. How does the behavior of the water around the drop compare with the behavior of light as it spreads out from a light source? Write your response in complete sentences under Step 4.

7 Your teacher will demonstrate how to generate waves using the wooden dowel. Practice generating waves (see Figure 12.8) by making:

    A. a single wave that travels the length of the tray.

    B. a set of waves that are as straight as possible.

    C. five or six closely spaced waves that travel the length of the tray at one time.

8 Observe the waves on the sheet of paper below the tray.

    A. What do you notice about the speed of the waves? Record your observations under Step 5 on the student sheet.

    B. What happens to the distance between the waves when you increase the rate at which you generate them? Write your answer under Step 6.

▶ PRACTICE GENERATING WAVES. TRY TO MAKE THE WAVES AS STRAIGHT AS POSSIBLE. SINGLE WAVES CAN BE GENERATED BY SLOWLY ROLLING THE DOWEL BACKWARD AND FORWARD A FEW MILLIMETERS (MM) IN THE WATER. MAKE CLOSELY SPACED WAVES BY RAPIDLY ROLLING THE DOWEL A FEW MM BACKWARD AND FORWARD. THE WAVES CAN BE BEST OBSERVED BY LOOKING BELOW THE TRAY WHERE THEY ARE PROJECTED ONTO THE PAPER. DO IT GENTLY. DON'T SPLASH WATER OUT OF THE TRAY.
FIGURE **12.8**

9 To model reflection, place the metal barrier in the ripple tank as shown in Figure 12.9. Direct waves at the barrier. Working with your group, use the ripple tank to model the reflection of light off a plane mirror. Record your procedure, data, observations, conclusions, and comments about the model.

10 Review your procedures, observations, and conclusions as you did in Inquiry 12.1 by using the discussion questions under Step 12 in that inquiry.

11 Share the results of your modeling with the class.

12 Dismantle your ripple tank. Pour the water out of the tray. Dry the tray with a paper towel. Return all the materials as directed by your teacher.

▶ PLACE THE METAL BARRIER IN THE RIPPLE TANK. DIRECT WAVES AT THE BARRIER.
FIGURE **12.9**

READING SELECTION

BUILDING YOUR UNDERSTANDING

WAVE OR PARTICLE MODELS

Many of the characteristics of light can be modeled as waves and as particles. Scientists find both models useful in understanding the behavior of light. These models allow scientists to discuss and think about light as electromagnetic wave energy or as particles (more accurately, as packets of energy called photons). Often, scientists use both of these ideas in thinking about the nature of light. ■

REFLECTING
ON WHAT
YOU'VE DONE

1. With your group, discuss how the behavior of ball bearings and water waves compares with the behavior of light.

   A. Draw a 3-column table in your notebook. Label the headings "Behavior of Light," "Behavior of Ball Bearings," and "Behavior of Water Waves." Use the table to summarize your ideas.

   B. From your observations, do you think light behaves like waves, like particles, or sometimes like both waves and particles? Record your own ideas about this. Be prepared to share your ideas with the class.

2. Read "Seeing Waves" on pages 157-159. Make a list of examples of waves described in the reading selection. Add some examples of your own.

SEEING WAVES

When most people think of waves, they think of waves in water. For example, you have probably seen waves breaking on a beach or waves made by someone jumping into a swimming pool.

Think about the waves produced when you throw a stone into a pool of still water. When the stone hits the water, you see a splash. Then a wave, or waves, radiates out from the point of impact. Where did the energy come from to make that wave?

All waves carry energy. A surfer riding a giant roller off the coast of California is using energy that may have been carried thousands of kilometers across the ocean—energy gained from the wind of a storm on the shores of another continent. As a wave moves through the ocean, it does not carry water with it. Instead, the water moves up and down in a circular motion as the wave passes through it. When the wave meets a distant shore, it breaks, transferring its energy to the shore. Sometimes this causes disastrous results. The large waves of hurricanes and other storms often destroy waterfront property.

▶ ALL WAVES TRANSMIT ENERGY. THE TRANSFER OF ENERGY FROM OCEAN WAVES TO THE SHORE CAN BE VERY DESTRUCTIVE.

PHOTO: Providence Journal Co./NOAA's National Weather Service (NWS) Collection

▶ MUSIC FROM THIS ORCHESTRA IS TRANSMITTED AS SOUND WAVES THROUGH THE AIR TO THE EARS OF THE PEOPLE IN THE AUDIENCE.

PHOTO: Craig Hatfield/creativecommons.org

WAVES ARE ALL AROUND US

You may not recognize them, but you see waves all the time. Think about flags flying on a windy day. If you look closely at the way they flap in the wind, you can usually see waves passing along the fabric. The energy carried by the wave comes from the wind and travels along the cloth.

Other solid objects also show wave motion. Look at a field of grass on a windy day. Energy from the wind creates waves that are transmitted from plant to plant as they sway. The top of the grass creates the illusion of a "sea of grass." Tall buildings and bridges also sway in the wind as waves are transmitted through their structures. Earthquakes are transmitted by waves that travel through the ground. They can make buildings sway so much that they collapse.

Some waves are invisible. For example, sound is transmitted by waves that travel through the air or other matter. We hear sounds when the waves strike our eardrums and cause waves, or vibrations, in the eardrums.

The types of waves that have already been discussed are called mechanical waves. Mechanical waves can travel only through matter (that is, solids, liquids, and gases). Light also can travel through some matter—through glass or air, for example. Light is considered to be a type of wave, but not a mechanical wave. Light is an electromagnetic wave.

Unlike mechanical waves, electromagnetic waves can travel through a vacuum—the absence of matter. This means they can travel through the emptiness of space. All waves—both mechanical and electromagnetic—transmit energy and have certain other features in common. ■

▶ **WAVES ARE TRAVELING ALONG THESE FLAGS.**

PHOTO: Dan McKay/ creativecommons.org

DISCUSSION QUESTIONS

1. What evidence can you present that demonstrates that waves carry energy?

2. Think about waves that you come in contact with during the course of the day and classify those waves as mechanical or electromagnetic.

The Greatest Scientific Argumen of the Millennium?

Scientists often debate ideas or theories for long periods of time. Sometimes more than one theory fits the facts. This is the case with the nature of light. The result has been a reasoned argument and debate that has lasted over 200 years.

PARTICLES OR WAVES?

Until the 1670s, people were very confused about what light was. At about that time, two scientists came up with different models to describe and explain light. You have already heard of one of these scientists—Isaac Newton (1642-1727). Newton suggested that light consisted of streams of particles moving at very high speeds. He called these particles "corpuscles" and used this model to explain the fact that light travels in straight lines. Newton's model could explain how light formed shadows, bounced off mirrors, and shined through the vacuum of space. He suggested that the different colors that make up white light were different particles.

For Newton to be able to explain refraction in terms of particles, light would have to travel faster in water or glass than it does in air. (We now know that light passing from air to water or glass slows down.) However, as you have read, at that time nobody could accurately determine the speed of light in air or in any other transparent material. So nobody could tell if Newton's theory best explained refraction.

At about the same time, another scientist, Christiaan Huygens (1629-1695), had a different explanation. Huygens suggested that light moved like waves traveling across the surface of a pond. He explained color by suggesting that each color was a different wavelength. According to Huygens, refraction occurred because light traveled slower in transparent materials, such as water or glass, than it did in air.

Huygens's theory of the nature of light was quite different from Newton's explanation. Both theories, however, could explain all the experimental data on refraction that was then available. As new discoveries were made and new data became available, whose theory would ultimately be correct?

Most scientists preferred Newton's idea. Some were betting on his reputation as the greatest scientist of all time. Others couldn't see how Huygens's light waves could travel through a vacuum. After all, how could you have waves in nothing? For more than a century, Newton's corpuscles were in and Huygens's waves were out.

Then along came Thomas Young (1773-1829) with a new type of experiment. In 1801, Young discovered that it was possible for two beams of light to interfere with each other and result in different colors or even darkness. He used this idea to explain why transparent materials such as oil films and soap bubbles often look multicolored.

Young called this process "interference" and explained it in terms of "out-of-step" waves interacting with each other. That is, light waves (or other waves) traveling in the same or different directions in the same space disturbed each other. For example, the crest of one wave would cancel or partly cancel the trough of another wave. Young theorized that this activity apparently resulted in some colors being enhanced and others being cancelled out altogether.

Other new discoveries were beginning to make waves look like a better model for light. Many invisible forms of light were being found, all of which could be explained as different wavelengths in the continuous electromagnetic spectrum.

▶ WHY DO TRANSPARENT SOAP BUBBLES DISPLAY THESE COLORS? THOMAS YOUNG EXPLAINED THIS PHENOMENON IN TERMS OF LIGHT WAVES INTERFERING WITH EACH OTHER.

PHOTO: By Jurvetson (flickr)

▶ MAX PLANCK SUGGESTED THAT ALTHOUGH LIGHT HAD MANY OF THE CHARACTERISTICS OF WAVES IT ALSO BEHAVED LIKE PACKETS OF ENERGY. WERE THESE PACKETS OF ENERGY THE SAME AS ISAAC NEWTON'S CORPUSCLES?

PHOTO: Library of Congress, Prints & Photographs Division, LC–B2-1250-11

Eventually, waves became the number-one theory. Had Newton at last been proven wrong? Was the debate over? That would be too simple! At the turn of the 20th century, German scientist Max Planck (1858–1947) was studying how hot objects gave off electromagnetic radiation. He discovered that these hot objects did not give off light energy continuously, but rather as packets of specific amounts of light energy (rather like the difference between integer and real numbers). He called these packets of energy "quanta."

This new evidence stirred up the debate even more, and attracted some of the world's greatest scientific minds. One of these scientists was Albert Einstein (1879–1955). Einstein used the idea of light as packets of energy to explain how light can knock electrons off the surface of some metals to produce an electric current. He called these packets of light energy "photons." Were these the same as Newton's corpuscles? Not exactly.

THE DUAL NATURE OF LIGHT REVEALED

Einstein was not reviving Newton's 200-year-old idea. Instead, he was suggesting that light behaved like both particles and waves. He described light as being waves that come in discrete packages, each containing a fixed amount of energy. This theory explained the existing data better because it clarified why light sometimes behaved like waves and sometimes like particles. Light had dual characteristics— some particle-like and some wave-like. The riddle of the nature of light had taken a new turn, and neither Newton nor Huygens had won or lost the debate. ■

NEWTON (LEFT) SUGGESTED THAT LIGHT CONSISTED OF STREAMS OF MOVING PARTICLES. HUYGENS (RIGHT) TOOK THE VIEW THAT LIGHT MOVED LIKE WAVES. LATER, SCIENTISTS LIKE EINSTEIN (CENTER) HAVE SUGGESTED THAT LIGHT HAS CHARACTERISTICS OF BOTH.

 DISCUSSION QUESTIONS

1. What characteristics of light does the wave model explain well? What characteristics of light does the particle model explain well?

2. Research another long debate in science and think about how it was resolved (or not!).

ASSESSMENT—HOW FAR HAVE WE COME?

SCIENTIFIC INVESTIGATIONS OFTEN INVOLVE CAREFUL MEASUREMENT AND RECORDING OF DATA.

PHOTO: Jeff McAdams, Photographer, Courtesy of Carolina Biological Supply Company

INTRODUCTION

This lesson is the assessment for *Exploring the Nature of Light*. The assessment has two sections. In Section A, you will conduct an inquiry into the size of shadows on a screen. You will take measurements and collect, record, and interpret data. Section B consists of questions, both multiple-choice and short-response. Some of these require you to use your knowledge and skills to interpret diagrams, data tables, and experiments. Your teacher will tell you when to do each section of the assessment and how long you will have for each section. In Inquiry 13.1, your work will be assessed partly on your layout, labeling, and drawing of data tables and diagrams. You and your teacher will use the results of this assessment to evaluate how well you can apply the knowledge, concepts, and skills you have acquired in this unit.

OBJECTIVES FOR THIS LESSON

Review the concepts and skills learned in this unit.

Conduct an inquiry into the size of shadows produced on a screen.

Record and interpret data.

Use knowledge and skills acquired to answer questions that relate to *Exploring the Nature of Light*.

▶ **MATERIALS FOR LESSON 13**

For you

1	copy of Student Sheet 13.1: Assessment Review
1	copy of Student Sheet 13.2: Sample Assessment Questions
1	copy of Student Sheet 13.3: Section A—Performance Assessment
1	copy of Student Sheet 13.4: Section B—Written Assessment Question Sheet
1	copy of Student Sheet 13.5: Section B—Written Assessment Answer Sheet

(Your teacher will explain whether the materials listed below are for you or for you and your group)

1	assembled light stand
1	white screen
2	plastic stands
1	black paper square attached to a craft stick
1	meterstick
1	metric ruler
1	sheet of white paper
2	binder clips

GETTING STARTED

1 With your group, review the concepts and skills described on Student Sheet 13.1. As you review the concepts, think about what you did in the inquiries in this unit and the evidence you have to support what you have learned.

2 After you have reviewed the concepts and skills, complete the sample assessment questions on Student Sheet 13.2. These are designed to help you understand the kinds of questions you will complete on the assessment. Your teacher will reveal the correct responses after you have completed the questions.

▶ IN THIS FINAL ASSESSMENT, YOU WILL CONDUCT AN INQUIRY INTO THE SIZE OF SHADOWS ON A SCREEN AND APPLY WHAT YOU HAVE LEARNED ABOUT THE NATURE OF LIGHT.

PHOTO: D. Sharon Pruitt/creativecommons.org

INQUIRY 13.1

MEASURING SHADOWS

SECTION A
PERFORMANCE ASSESSMENT

PROCEDURE

NOTE Read all instructions before you start working. Record all your responses on Student Sheet 13.3: Section A—Performance Assessment.

SAFETY TIP

Do not touch the lightbulb in the light stand. It gets very hot and may cause painful burns.

1 Set up the light stand so that the lightbulb filament is horizontal.

2 Place the screen 50 cm from the end of the lightbulb filament.

3 Use the binder clips to attach the white paper to the white screen.

4 Use the black square (attached to the stick) to produce a shadow on the paper on the screen (see Figure 13.1).

Meterstick

50 cm mark

30 cm ruler

Binder clips attach the white paper to the screen

▶ USE THIS APPARATUS TO FIND THE AREA OF THE SHADOW YOU PRODUCE ON THE PAPER.
FIGURE **13.1**

Section A continued

REFLECTING
ON WHAT
YOU'VE DONE

5 Measure the size of your shadow (in cm) when the black square is 5 cm from the screen. On Student Sheet 13.3, record the dimensions of the shadows, and calculate the area of the shadow (width × height) in square centimeters.

❶ Discuss the performance assessment with the class and identify wh concepts and skills v applied.

6 Repeat this procedure, measuring the size of the shadow at four additional distances farther from the screen.

❷ Review the correct responses for the multiple-choice and short-response questions. Discuss th responses, and ask questions to clarify y understanding of an concepts.

A. Record all five sets of measurements of the distance of the black paper square from the screen, shadow dimensions, and shadow area in a data table of your own design. Create your table under Step 3 on your student sheet.

B. What happens to the size of the shadow as the square is moved farther from the screen? Record what you observe under Step 4 on the student sheet.

C. Use words and a labeled diagram(s) to explain your observations under Step 5.

SECTION B
WRITTEN ASSESSMENT

PROCEDURE

1 Your teacher will outline the procedure for taking Section B of the assessment. Work individually. Do not write on Student Sheet 13.4: Section B—Written Assessment Question Sheet. Write only on Student Sheet 13.5: Section B—Written Assessment Answer Sheet. Hand in all papers before you leave the class.

Glossary

absorb: To take in and not give out again.

additive color mixing: The type of color mixing that takes place when colored lights are mixed together.

amplitude: The distance from the midpoint of a wave to the crest (or trough) of the wave.

angle of incidence: The angle at which a ray of light strikes the surface of an object, and which is measured from the normal. See also *normal.*

angle of reflection: The angle at which a ray of light reflects off the surface of an object, and which is measured from the normal.

apparent brightness: How bright an object appears.

Celsius: A temperature scale with the melting point of ice at 0°C and the boiling point of water at 100°C (at standard atmospheric pressure).

chemical energy: The form of energy stored in substances such as food and fuel that is transformed to other types of energy during chemical reactions, for example, when gasoline is burned.

chemical reaction: Any change (other than a nuclear reaction) that involves the formation of a new substance.

chlorophyll: A green substance found in the chloroplasts within the leaf cells of green plants. It plays an important role in photosynthesis, in which light energy is transformed to chemical energy, which is then used to make carbohydrates (food).

chloroplasts: Microscopic structures containing chlorophyll that are found in some plant cells.

color: A visual sensation produced by different wavelengths of visible light.

color filter: A transparent colored object that transmits some wavelengths of visible light while absorbing others.

component: A part of something.

crest: The highest point of a wave. See also *trough.*

diffuse light source: A light source that emits light from more than one point on its surface, such as the Sun or a standard light bulb. See also *light source.*

diode: An electronic component, some forms of which—for example, light emitting diodes (LEDs)—can release light.

eclipse: The passage of an object through the shadow of another object. The term eclipse is usually applied to astronomical objects, for example, planets or moons.

electrical energy: The form of energy associated with electric charge and electric current.

electromagnetic radiation: Energy (contained in electric and magnetic fields) that can be transmitted without matter and may be thought of as waves or particles (photons). It includes visible light. See also *electromagnetic spectrum*.

electromagnetic spectrum: The range of wavelengths (from radio waves to gamma waves) over which electromagnetic radiation extends. See also *electromagnetic radiation*.

electron: A subatomic particle that carries a negative electrical charge.

element: A substance that cannot be broken down into other substances by chemical or physical means (except by nuclear reaction).

energy: The ability to make things do work.

energy transformation: The change of one form of energy into another (for example, electrical energy into light energy).

Fahrenheit: A temperature scale with the melting point of ice at 32°F and the boiling point of water at 212°F (at standard atmospheric pressure).

filament: The part of a lightbulb that releases heat and light when an electric current is passed through it.

frequency: The number of waves passing a point in a given time.

hertz: The number of waves passing a point each second.

incident ray: A ray of light striking a surface.

infrared: Wavelengths of invisible electromagnetic radiation that are slightly longer than those of visible red light.

interference: A disturbance that may occur when waves (including light waves) travel in the same or different directions in the same space.

light source: An object that produces light.

light year: The distance light travels through a vacuum in one year.

luminous: Producing or emitting light.

megahertz: A unit of frequency equivalent to 1 million wavelengths per second. See also *hertz*.

normal: An imaginary line that is perpendicular to the surface of an object. The normal can be defined for any point on a surface. See also *angle of incidence*; *angle of reflection*; *refraction*.

opaque: Not allowing any light to pass through.

optics: The study of light.

particle: The smallest amount of matter or energy of a certain type that can exist. For light, this is a photon. See also *photon*.

penumbra: An area of semishadow formed around the umbra that can be produced only by a diffuse or large light source. See also *umbra*.

perceive: To sense something, as in the way sense organs and the brain work together to make us aware of something. For example, the eye and the brain work together to recognize wavelengths of light, and mixtures of wavelengths, as color. The process of perceiving is called perception.

photon: The smallest energy packet of light (quanta) or other electromagnetic radiation of a given wavelength that can exist.

photosynthesis: The process by which plants make food (glucose) and oxygen from water and carbon dioxide, using light as the energy source for this chemical reaction.

pixel: A small dot that makes up part of an image (for example, on a TV screen).

plane mirror: A flat mirror.

point light source: A light source that emits light from one point (or from a very small area). See also *diffuse light source; light source*.

primary colors: One of three colors that can be combined to make up any desired color. For colored lights these are red, green, and blue, which are different than the primary colors used in paint or ink color mixing (cyan-blue; magenta-red; and yellow).

prism: A transparent object with at least two nonparallel flat sides.

quanta: The smallest packets of energy possible. Quanta of light are called photons. The singular form of quanta is "quantum."

ray: The path of light as it travels. Rays are represented as lines with arrows that show the direction in which the light travels.

reflect: To change the direction of a wave when it bounces off a surface.

reflected ray: A ray of light bouncing off a surface.

refract: To change direction or bend from the original path. Light refracts (changes its direction of travel) when it passes from one medium into another medium of different density.

refraction: The bending of light as it passes from one medium into another.

scatter: To absorb and then emit light in random directions.

shadow: An area of darkness that forms behind an object when the object blocks a source of light.

spectrum: A distribution of colors or wavelengths of electromagnetic radiation. The plural form of spectrum is "spectra."

speed: The distance traveled in a specific time (distance divided by time).

subtractive color mixing: The type of color mixing that occurs when colors are removed (for example, by a color filter) from mixtures of light.

total internal reflection: Reflection of all the light at the boundary between two transparent substances.

translucent: Allowing light to pass through, but scattering it (for example, frosted glass). See also *opaque*; *transparent*.

transmit: To allow the passage of energy or matter. For example, a red color filter allows red light to pass through it, or transmits red light.

transparent: Allowing light of certain wavelengths to pass through without significant scattering. See also *opaque*; *translucent*.

triangular prism: A transparent object with three sides and flat ends that can be used to split white light.

trough: The lowest point of a wave. See also *crest*.

ultraviolet: Wavelengths of invisible electromagnetic radiation that are slightly shorter than those of visible violet light.

umbra: The inner, dark region of a shadow. No portion of the light source can be seen from within the umbra. See also *penumbra*.

visible spectrum: The distribution of colors produced when white light is split.

wave: A disturbance that moves energy through matter or space without carrying matter with it.

wavelength: The distance from one wave crest to the next (also the distance from one wave trough to the next).

white light: A mixture of different colors of light that we detect as white or colorless.

Index

Photo Credits

Front Cover
NASA Jet Propulsion Laboratory

Lessons
2 Amy Snyder, © Exploratorium, www.
exploratorium.edu 9 dougwoods/
creativecommons.org 10 (top) Tony Hisgett/
creativecommons.org (bottom) mattgarber/
creativecommons.org 11 Courtesy of Gary
Berdeaux 12 (left) Library of Congress, Prints
& Photographs Division, LC-USZ62-111797
(right) Library of Congress, Prints &
Photographs Division, LC-USZ62-60242
13 (top) DoD photo by Petty Officer 2nd
Class Ryan C. McGinley, U.S. Navy (bottom)
Clinton Steeds/creativecommons.org
14 Nomad Tales/creativecommons.org
15 (top) javajones/creativecommons.org
(bottom) Nomad Tales/creativecommons.
org 16 Library of Congress, Prints &
Photographs Division, LC-DIG-jpd-01968
17 NOAA 18 DoD photo by A1C Isaac
G.L. Freeman, USAF 20 jennifer chong/
creativecommons.org 23 Josep Mª Rosell/
creativecommons.org 26 (top right) Clark
Gregor/creativecommons.org (bottom)
Jeff McAdams, Photographer, Courtesy
of Carolina Biological Supply Company
27 kcdsTM/creativecommons.org 28 Erik
Charlton/creativecommons.org 29 (top)
ItzaFineDay/creativecommons.org (bottom)
Courtesy of DOE/NREL, Credit—Coherent
Inc., Laser Group 30 OAR/National Undersea
Research Program (NURP) 31 (top left) Doug
Myerscough (bottom right) James E. Lloyd,
Department of Entomology and Nematology,
University of Florida, Gainesville 32 NASA

38 NASA/Johns Hopkins Applied Physics
Laboratory/Southwest Research Institute/
Goddard Space Flight Center 39 (top left)
Library of Congress, Prints & Photographs
Division, LC-USZ62-7923 (bottom left)
Courtesy of the Smithsonian Institution
Libraries, Dibner Library of the History of
Science and Technology, Washington, D.C.
(top right) Library of Congress, Prints &
Photographs Division, LC-USZ62-124161
(bottom right) Courtesy of Smithsonian
Institution Libraries, Dibner Library of
the History of Science and Technology,
Washington, D.C. 40 NASA/JPL/California
Institute of Technology 44 Jason Pratt/
creativecommons.org 46 By Jurvetson
(flickr) 48 (top right) OAR/National Undersea
Research Program (NURP); Harbor Branch
Oceanographic Institute (bottom left) NOAA
Photo Library, NOAA Central Library; OAR/
ERL/National Severe Storms Laboratory
(NSSL) 49 © 2009 Carolina Biological
Supply Company 52 NASA 54 (top) Photos
of Javanese shadow puppets from the
private collection of Tamara Fielding and
Tamara and the Shadow Theatre of Java;
www.indonesianshadowplay.com (bottom)
Photos of Javanese shadow puppets from
the private collection of Tamara Fielding
and Tamara and the Shadow Theatre
of Java; www.indonesianshadowplay.
com 56 Ctd 2005/creativecommons.org
58 Nancy Rodger, ©Exploratorium, www.
exploratorium.edu 63 (top) Courtesy of
Carolina Biological Supply Company
(bottom) © M.J.F. Marsland 2000
64 Rhett Maxwell/creativecommons.org